Humane Physics – The Whole Story

(No student was harmed in writing this book)

by

Francis Mont

A conceptual biography of Fundamental Ideas

for Vera

my brilliant editor
who listened to me with almost infinite patience

Published by Montland Books in 2020
Third Edition
ISBN: 978-0-9959174-9-1
Cover Photo by:
© Fantasista | Dreamstime.co

m

Contents

Book I – Classical Physics

Introduction

The Author

The first event in my life that suggested that I would become a physicist happened when I was eight years old. My big brother was already building radios, and I asked him to teach me. He explained the components. He got as far as 'condenser'. When I asked him what it was, he started telling me about electric charges, but when I asked him what *they* were, he had to backtrack and tell me about electrons. When I demanded to know what electrons were, he gave up on me. I never built the radio.

I became interested in science in high school when I realized I could actually use it in my own life. Two incidents stand out.

A very deep cave on the hill near my home fascinated us, kids when I was sixteen. The cave went vertically down, impossible to climb into, and the neighborhood boys made wild guesses about its depth. Days after we learned the Newtonian equations of gravitational free-fall, I put an end to the guessing game. All I had to do was drop a stone and count how long it took to strike the bottom. Three seconds – about 44 meters. I was much admired by my classmates. Naturally, that success made a deep impression on me.

A couple of years later, I wanted to go on a hiking trip with two friends, and we needed a tent. My father had a huge old army tent, badly damaged by mildew. Large parts of it were still in good condition, so we decided to make a new tent from the sound fabric. I used fairly complicated trigonometry and calculus to figure out the maximum size of the tent we could make from the available irregularly shaped material, providing a large enough floor space for three to sleep on and sufficient height for getting in and out. When we started cutting, we weren't sure whether it would work. After all, it was only math. It worked! We made a beautiful new tent out of an old wreck. After that, I was sold on science.

8   Humane Physics

Intending to become a science teacher, I enrolled in the University of Science in Budapest. I was working on my Teacher's Diploma in Mathematics and Physics, with only two years left, when Einstein ruined everything. He started telling me how time is relative and space is warped and how I can take a trip of a few weeks and come back hundreds of years later. This was all utter nonsense. Still, Einstein himself I couldn't entirely dismiss, I had to deal with these concepts. So, I transferred from the teacher's program to a graduate course in theoretical physics and set out to prove Einstein wrong. Or at least find out what the hell he was talking about.

I didn't like the way they taught us physics. The university aimed to train us to be useful members of the workforce: competent in calculating parameters of electronic circuits, of stresses in bridges and tall buildings, of the speed of semiconductor switches in computers. The purpose of the graduate course was definitely not to teach us how the world is put together and what is the essence of basic principles. I found this out, too late after I graduated.

My ambitions were modest: first, to develop my Unified Field Theory (a single equation explaining the universe); then, to provide the theoretical foundation for an anti-gravity device that would revolutionize science and technology, incidentally creating Heaven on Earth.

With the ink still wet on my diploma, I stood in the job placement center, talking to an increasingly frustrating counselor. Though I had already told him what I wanted, he was obstructive.

"I have here a position in the light-bulb factory. You could work on developing a longer-lasting filament," he suggested.

"No", I said firmly.

"Well, how about developing a better-insulated transistor casing", he tried again. "I have two positions there".

"No," I said, starting to lose patience.

"The Geophysical Institute is looking for physicists" he offered. "They provide rubber boots for..."

I snorted.

"I've got it!" he exclaimed. "A good position in the cheese factory. They need someone to measure viscosity."

I said nothing. He glanced at my face and quickly looked away. After a while, he said very quietly, making sure nobody heard: "You could join the first computer company in Budapest. They are looking for anybody with a scientific background. You wouldn't do physics, but – he looked around even more carefully, lowered his voice even further, and leaned quite close to my ear - this could be your ticket out. Sooner or later they would have to send you abroad. In the West, you could research whatever you liked".

So, I did and arrived in Canada a year later. I have been having fun with both physics and computers ever since.

I have always wanted to write a book on physics, concentrating on the basic laws of nature, the major breakthroughs in our understanding, and the scientists who contributed most to our knowledge. The book became suddenly urgent when I realized that the world around me had started to lurch in a very dangerous direction after September 11, 2001. What I found most frightening in all that has happened since was the ease with which political leaders exploit people's fears and lead them on a course that would have been unthinkable even a few years before.

Suddenly, due process of law was discarded; torture became a legitimate tool used by the government; unprovoked military aggression against a weak but oil-rich country was generally accepted. Intimidation of, and spying on, citizens became the uncontested norm, and billions and billions of dollars were taken out of social programs and disappeared into the bank accounts of large corporations that line up for rich military contracts.

I could not help wondering how it was possible - how people could be so blind and gullible. I found the answer in a steadily deteriorating education system that never taught young people how to think logically, critically, scientifically, and in a cultural environment that focuses on shallow distractions. Then the idea hit me: if we all understood the basic rules of logical thinking, the inevitable steps we have to follow to learn truth and reality, the same rules and steps scientists have had to follow every day of their productive lives – then deception and exploitation of the citizenry would not be possible.

I want to show what science means to humanity; how it has always represented the best in us: curiosity, observation, intelligence, creativity, original thinking, cooperation, and dedication.

The book is also a history of physics, concentrating on fundamental principles and major breakthroughs. It is amazing how few really basic concepts the entire body of science and technology depends on. For every major discovery, there have been thousands and thousands of smaller advances, innovations, and applications. I followed this line of insights from ancient Greece to Stephen Hawking. It can be used as an introductory or supplementary text on physics. I have tried to explain the main concepts in the way I would have liked them explained to me when I was at university. I still remember the problems I had understanding the essence of these concepts and principles and I would like to share the results of my decades-long search for clarity with both students and the interested public.

All through the book, I was trying to show people the best examples of creative, critical problem-solving techniques that could and should be used in our social and political lives. We have to learn to think independently, or humanity is doomed: it will fall prey to exploiters who thrive on complacency, inertia, and credulity.

I hope you have as much pleasure in the reading as I had in the writing of it.

The Book

In recent years, more and more books have been published on science in general and physics in particular. Yet I notice a curious gap. Each book focuses on only one or two of the following aspects of physics: history, biography, theory, methods, applications, and social significance.

I am not aware of any work that tries to give the reader a rounded picture, dealing with all of these topics in one book, in an integrated, comprehensive way. This book is intended to fill the gap; to teach the fundamental principles of physics. The lives and significance of the principal scientists are summarized, non-mathematical explanations of basic concepts and fundamental laws are provided and consequences and applications are discussed.

For advanced readers, the "Next Level" chapter points the way for further study, using advanced high-school mathematics. The required math is summarized in the short "Your Math Toolkit" chapter just ahead of the "Next Level". Many of the results presented in the main body of the book are mathematically proven and some typical examples are calculated and solved there.

Laymen, high-school, and college students could all benefit from a book that gives them a birds-eye view of all aspects of the science of physics in a precise, fundamental and integrated way. I sure missed this kind of book myself, both as a student and as a teacher.

If you are looking for a regular Physics textbook, with lots of diagrams, examples, detailed descriptions of experiments, and practical problems to solve at end of each chapter, this book is not for you. At least not as primary source material. You can use this book as a supplementary source if you want to understand the deeper significance of the concepts and discoveries, but you may be better off starting with a standard textbook. For those who want that kind of Physics, I can highly recommend Sheldon Glashow's magnificent work: *From Alchemy to Quarks*.

This book was written for those who want to understand physics, in the human context. What I attempted to do was follow the basic discoveries through the minds of the scientists and try to emphasize the essence of these in a larger philosophical and social context, both in their time and today.

Many people are reluctant to read a book on physics because they think the math is beyond them. That is why I wrote the book in two sections: the first dealing with the human context: theory, principles, history, biography, significance. This part has practically no math in it. The second, what I called the "Next Level", is for those who want to know what the rest of physics looks like, as practiced by thousands of physicists around the globe, every day. For that, you need mathematics and to follow its rules without hesitation, regardless of how startling the results turn out to be. The only question is whether reality agrees with the consequences and predictions of your math. As Niels Bohr told Wolfgang Pauli during a conference: "We are all agreed that your theory is crazy. The question that divides us is whether it is crazy enough to have a chance of being correct." Once you understand this, you will understand the deepest essence of physics and physicists.

Physicists are just like the rest of us, and the stories of their motivation, successes, and failures will put us in their shoes as we follow their progress from confusion to clarity. When you read about them, try to imagine what it would be like meeting them in person, asking them questions, listening to their concerns over the mystery and the confusion. They are all fascinating characters with one thing in common: a deep desire to find things out, to see the light of truth, to be the first to get there.

The story of physics is not a straight line from concept to discovery. It is a tortuous journey with many setbacks, repetitions, discoveries, and re-discoveries, often going around in circles. Many of the physicists struggled with the same mysteries, fighting among themselves, often passionately, over minor differences in interpretation. The result of this all too human process is our current technological civilization with both its benefits and dangers for our survival. That is why I chose the title of "Humane Physics".

One last comment: throughout the book I used the male pronoun. Not because I am sexist (far from it), but because in many languages, including my own mother tongue, there is only one third-person pronoun and it is used for both sexes. I find the he/she or (s)he compromise awkward, as it interrupts the flow of thought, and 'they' can be confusing. I decided to use what is most familiar to me. I hope no one will be offended.

<u>Saturn</u>

The pale, pastel-pink globe,
embraced by jaunty rings,
on a vast, star-studded sky,
drifted, slowly,
across my viewfinder.

…Earth slowly rotated
under my feet,
summer-dawn breeze
gently ruffled my hair…

Saturn *was* there, immutable,
following its path of causes and effects,
and I thought of Galileo,
who laboured over a block of glass,
grinding it into a lens,
to behold the same planet…

…and, through a gap of four hundred years,
he and I
looked into each other's eyes.

Chapter I – Science

How big is the world?

Whenever I start talking about physics, people's eyes glaze over. They remember from science class all those dull experiments with pulleys and springs and little carts rolling along on the teacher's desk. Or they think of the totally confusing and incomprehensible ideas of Einstein and his relativity, the "twin paradox" and the curvature of space ("How can space, which is a nothing, be curved?" – rolling eyeballs).

I think this aversion could be overcome by starting with astronomy, as the study of physics actually began in pre-history. The teacher should lead the class out to an open space, on a clear, starry night and show them the moon through a telescope: a huge marble ball, hanging there in empty space, unsupported, slowly moving across the viewfinder. See the craters and the sharp line separating the illuminated surface from the part in shadow but still visible. Or find Venus and observe its current phase, or Saturn with its magnificent rings or Jupiter with its moons. Look at these planets as they *move*, forcing you to make frequent adjustments to keep them in view. And then aim your telescope at a star and observe how it floats there, unmoving, steady, mysterious.

If the teacher has access to a more powerful telescope, he should aim it at the constellation Draco (between the Big Dipper and Little Dipper) and following its tail, find the brightest pair of galaxies in the sky: M81 and M82, with their magnificent spiral arms: a rotation frozen in the fleeting moment of a human life-span. Then the student might feel as Richard Feynman must have when he wrote in his *Lectures on Physics*: "If one cannot see gravitation acting here, he has no soul."

I still remember the awe and wonder I felt as a child, looking up at a star-studded sky and wondering what they were like close up, how many planets there had to be around those stars, maybe with an intelligent life-form, looking 'down' at me right now, thinking the same.

I knew then that I had to learn everything I could about this wonderful universe.

And now, we are ready to do it together.

……………………….

Once, many years ago, my son asked: "Daddy, how big is the world?"

When I told him that Earth is a globe of about 12,000 km in diameter or 40,000 km around the middle, he asked

"How do you know?"

When I told him that I had learned it in school and it is in all the atlases and textbooks he asked again, with the persistence of a seven-year-old: "But how did the people who wrote the books know?"

There was no point telling him that those authors had also learned it in school; I knew he would not let me off the hook. So, I promised to find out. Isaac Asimov's *Chronology of Science and Discovery*, Carl Sagan's *Cosmos,* and Simon Singh's book: *Big Bang* tell us the story:

Apparently, Eratosthenes, chief librarian of the Alexandrian library, in 240BC heard that in the city of Syene (now called Aswan) in Egypt, on June 21st, at exactly noon, sunlight reached the very bottom of a deep and narrow well.

He wondered whether it would be the same in Alexandria. On the next June 21st, he went outside the library and stuck a perfectly straight pole vertically into the ground. At high noon, he was surprised to see a definite shadow of the pole, which told him that the sun was not directly overhead, and thus, could not illuminate the bottom of a well where he lived - while a few hundred miles to the south, it did. The only logical explanation was that the Earth must be round, just as the full Moon appeared to be.

Once Eratosthenes had digested this portentous thought, he started wondering how big a sphere the Earth might be, and how he could find out. Knowing a fair bit of geometry, he realized that all he needed to do was to measure the distance from Alexandria to Syene and the angle of the sun's rays shining down on his pole at noon on June 21st. Measuring the angle was easy. To measure the distance between the two cities, he hired a man to walk to Syene, counting his steps.

From these two pieces of data, he calculated the circumference of our planet as 24,200 miles or 38,962 km. He must have had a very conscientious walker because this figure is very close to the value we can measure with high-tech instruments: 24,902 miles or 40,092 kilometers.

Then, he started wondering what else he could learn.

Looking up from Earth, the moon and the sun were the two most visible objects in the sky. How large were they?

Being the chief librarian of the Great Library of Alexandria has its advantages. With a little research, he learned that the relative size of the moon had been calculated 10 years earlier by Aristarchus, who used the lunar eclipse to accomplish this feat. Aristarchus measured the time it took the moon to completely disappear in the Earth's shadow (50 minutes) and the time it took to fully emerge again (200 minutes). From these two numbers, he determined that the moon's diameter had to be one-quarter that of the Earth. Since Eratosthenes had just calculated the Earth's diameter, he now knew that the diameter of the Moon had to be about 3200km.

The next piece of the puzzle was the distance of the moon from Earth. Being in possession of the diameter of the moon, it was easy to estimate its distance. He did this by holding up a thumb to cover the full moon, with his arm held straight out before him. He knew that the ratio of his thumb's diameter to the length of his arm - about 1:100 - had to be the same as the ratio of the moon's diameter to its distance from Earth. Consequently, the distance of the moon had to be 100 times the diameter of the moon or about 320,000 km. Today, we measure it as 384,400 km. The distance of the moon was calculated independently by Hipparchus in circa 150BC, using parallax - observing the moon's position against the stars as seen from Syene and Alexandria, respectively - and trigonometry, which he had invented.

Two more items could be determined by further observation and logic: the distance of the sun from the Earth and the diameter of the sun itself.

To find out the sun's distance, all Eratosthenes had to do was to measure the angle between the lines of sight to the moon and to the sun on a half-moon day, when both were visible. Knowing that on half-moon days the Earth, moon, and sun are on three vertices of a right-angled triangle (as suggested by Aristarchus), he could calculate the distance to the sun from the angle and the known distance to the moon. The result was very close to the one we know today: 150,000,000 km.

The last datum, the size of the sun, fell into place now, by the same reasoning he used to determine the distance to the moon – except now the moon took over the role of the thumb. During a solar eclipse, the moon exactly covers the sun. Therefore, the ratio of the moon's diameter to the moon's distance to the Earth is the same as the ratio of the sun's diameter to the sun's distance to the Earth. Since he now knew the sun's distance to the Earth, he was able to estimate the sun's diameter - the first approximation ever made.

Before I told my son these stories, I did a little more research. He was likely to ask about the stars next. I found out that the first person who measured a star's distance from Earth was Friedrich Bessel, a German astronomer, in 1810. He made very precise measurements of the position of a star called 61-Cygni. He actually made two sets of measurements, six months apart when the Earth was at opposite sides of its orbit around the sun. He noticed that the position of 61-Cygni shifted a very small amount between the two sets of observations and, using this minute shift and his knowledge of the diameter of Earth's orbit, he determined that the star had to be about a hundred trillion (100,000,000,000,000) km away. A large number indeed!

I was very proud of myself when I went back to my son and told him all these stories, but he immediately asked another question:

"Dad, how do we know that the Earth goes around the sun and not the other way around? I see it come up every morning from my window and see it go down through the other one in the living room."

That question sent me back to my books for more research. Nothing dramatic showed up, so I decided to tell him about how deceptive appearances are. Yes, it appears as if the sun moved around the Earth, but this is an illusion, due to the Earth's rotation around its axis. It also appears as if the Earth were flat, but after a few observations and some thought, we had to admit that the Earth is a sphere. I still remember my own childhood fascination with the thought that kids in China were walking upside down and didn't even know it. (Of course, there must have been a Chinese kid amused by thinking of *us* upside down).

The ancient Greeks noted that when a ship approached their harbour, it was spotted first as the top of a mast, then the sails, then gradually the hull emerged as the vessel got closer, Just as you would see a fly crawling over the surface of an orange. The Greeks had to accept that the surface of the ocean is curved. Similarly, learning about how different constellations came into view as one traveled north or south also suggested that our Earth was a sphere. Yet, it seemed incredible that "down" was different for people on different parts of the planet.

The chief observations that forced scientists to realize that all the planets, including Earth, move around the sun, were the planets' irregular motions. Unlike the fixed stars, these bodies move back and forth in a narrow band of the sky called the Zodiac lane, in a more or less irregular pattern, sometimes slowing down, stopping and turning back, then resuming their forward motion. Astronomers had been trying to explain this phenomenon for centuries, but only one theory was consistent with all the accumulated data: the irregular motion of the planets is due to the fact that we observe them from a moving platform. Like the Earth, with us aboard, travels around the sun at a speed different from that of other planets, sometimes we overtake them; thus, they fall behind and seem to turn back in their path.

My son would have to be content with this explanation because I had no further answers myself. We had arrived at the point where simple observations and logic would reveal no more truths about the universe. We had to learn the language of mathematics and methods of complicated and indirect reasoning to reveal Nature's secrets.

As a consolation for a universe more complex than a seven-year-old would like, I decided to show him some planets. We set up the telescope in our backyard. Finding Venus was easy. On a summer morning, before sunrise, it hangs low over the eastern horizon and is brighter than any star. Venus, in a phase where only half of the surface is visible, looks much like the moon.

Saturn is a bit harder to find; it is much dimmer than Venus, being so much farther away. The sight of that tiny globe hanging there, surrounded by its rings, was magnificent. It was actually there, floating in space and slowly drifting across the sky. I promised to show him Jupiter next, with its moons orbiting it, just as they were when Galileo discovered them four hundred years earlier. My son looked at me with big, round, bright eyes and said: "Dad, I want to know everything about the world. When I grow up, I'll be a scientist".

I hugged him - not because I took the promise too seriously, but because I was so relieved that he didn't ask any more questions.

Science

Science is a wonderful tool. We get home from a busy day at the office or the shop and flick the light switch as we enter our home, instantly turning night to day. Yet most of us, most of the time, don't think about what a miracle this is.

With that casual flick of our fingertip, we set into motion a very long chain of causes and effects, covering an entire continent, maintained by thousands of people who make sure every single component is where it's supposed to be. Should a connection be severed anywhere in that grid, our light may not turn on. Then we would have to grope our way in the dark to the telephone - another miracle - to report an outage.

Most of us almost never think of the very, very few people whose curiosity, thirst for knowledge, fascination with, and awe of, nature compelled them to tinker, experiment, try to figure out how the world is put together.

When we turn on the light, we don't think of Faraday, 183 years ago, wondering about Oersted's experiment showing how an electric current affected a magnetic needle. He wanted to know whether the reverse was also true: did magnets have anything to do with electric currents? So, Faraday twisted a piece of electric wire into a loop, connecting the two ends through an instrument that detects the flow of electric current.

That was the whole experimental setup that led to the light switch and made all our modern conveniences possible: power tools, the MRI, television, the microwave oven, DVD player. Everything that uses electricity became possible at the moment when Faraday picked up a bar magnet and waved it over the loop of wire… and the needle of the galvanometer moved.

The incredible implication of this story is that the ancient Greeks could have had electricity 2000 years earlier if the idea of experimenting had been popular then. They had all the requirements: wire (used in the making of jewellery) and magnets (the lodestone was studied by Thales around 585 BC). What they lacked was the mindset of a Faraday and his obsession for finding things out.

It may take a genius to discover and interpret the often very subtle signs we are given regarding the laws of nature but, once explained, they can be understood by anyone who makes the effort.

So, what is science? How does it work? What do scientists do? Why is it so difficult for so many people?

A poetic image

Imagine, if you will, that you are in an airplane, flying over the landscape of physics and technology. Below you are a few dozen workbenches scattered about, with curious little figures bending over them, moving artifacts here and there and back again - wires, beakers, jars, strange objects, and instruments. Behind them, at a respectful distance, an army of technicians and engineers waits impatiently. Every now and then something new happens on one of the front benches; a golden nugget falls to the ground and rolls away.

The engineers instantly swarm over it, picking it up, turning it this way and that, prodding, poking, molding it in different shapes, adding and improving and adapting it. And when they're done, out come things familiar to us today: light bulbs, telephones, electric motors, radio telescopes, washing machines, refrigerators, dentist drills, computers, the internet, street cars, electron microscopes, airplanes, and on and on and on.

Try to imagine a world without electricity! As recently as 135 years ago - Edison's first successful light bulb experiment was on October 22, 1879 - none of the things I listed above were available, most of them did not exist even in science fiction novels. People used candles, lanterns, coal and wood fire, horses and buggies, and back-breaking labor even for simple tasks like plowing a field or lifting a beam.

All that changed due to less than two dozen strange characters bending over their workbenches, muttering to themselves. Today we call them scientists and they number in many thousands. But at the beginning, only a few made the fundamental discoveries everything else was built on.

How did they do that? That is what this book is about.

The scientific method

Years ago, the chief engineer who hired me for a job told me the following joke:

"During the French Revolution a priest, a politician, and an engineer are sentenced to be executed by guillotine. Each is given a chance to make a statement before the execution is carried out.

The priest delivers a moving sermon about brotherhood, love, and peace. Then his neck is placed on the block and a string is pulled. The blade stops an inch above his neck. The crowd goes wild: a miracle happened, they let him go.

The statesman makes a fiery speech about patriotism, duty, and the rule of law. Neck on block, string pulled, blade stops an inch short. Another miracle – they let him go as well.

When it is the engineer's turn, instead of an address, he says to the executioner: "Unless you undo the knot in the rope, this damn thing will never work!"

The point my boss was trying to make was that he expected integrity from me: an attitude of wanting to get the job done, even if I could be hurt doing it. Rule number one in science: worship truth above all else!

The "Scientific Method" we scientists keep talking about has many ingredients, but this is the most important: intellectual integrity. Some call it *objectivity*: look at all the relevant data and go where the evidence takes you, whether you like it or not. Kepler discarded the results of years of very hard work because his theory did not quite agree with the observations. Einstein once published a paper documenting the colossal failure of research that had taken him two years, "to save another fool two years should he wander down the same path".

You might say that I am talking about plain honesty, not something special reserved for science and scientists. Yes and no. The difference is that scientists must take it seriously because without it the whole endeavor is pointless. No 'white lies' are possible in science. We cannot make an exception and suspend our objectivity, even for a second, because we are forging chains of logic and if even one link is missing or 'weak', the whole thing falls apart. We might fool ourselves, but we cannot fool Mother Nature, who doesn't care whether we manage to uncover her secrets.

Honesty and integrity are only prerequisites for science. They are not the method itself. The method starts with **observation**. Not just looking but seeing.

If you look at the stars on a clear night, your first impression is randomness. Thousands and thousands of pinpricks of light scattered all over the sky. After a while, you begin to see structure. The bright band of the Milky Way should be the first thing you notice. Then, if you have the time and imagination, you might look for some way to remember the shape of star groups. Maybe they remind you of something: a cart, a bear, a dragon…. Eventually, you might realize that the arrangement of stars within a group and the relation of the groups among themselves never change -- *except* for half a dozen 'stars' that seem to wander all over the place, moving forward for a while, then stopping and moving backward, then forward again.

It takes time to observe nature and find patterns that are not immediately obvious. Once you have observed something, you have to be able to describe it to others, which means you have to be able to describe it to yourself first. You give names to what you observed to save time in both thought and communication. The fixed lights, you call stars; the wandering few you call planets ('planetes' means 'wanderer' in Greek). Inventing names and explaining what they mean is called '*definition*'.

And that is where mere curiosity turns into science. Of course, scientists don't just sit down and make up definitions before doing anything else – it doesn't work that way. We must have some ideas about the attributes of matter and the processes that we want to study. Often, we understand something intuitively before we are able to give precise definitions. 'Force' and 'mass' in Newtonian mechanics were described before they were defined. Some concepts, like 'space' and 'time', we can never define precisely.

Science is not a linear progress, leading from 'A' to 'B'. It can be best described as 'iterative': getting closer and closer with every pass we make at it. We gain depth in the process and our definitions and statements will be more and more precise. A common error made by undergraduate students is the attempt to treat physics like mathematics. You can't because it isn't! Mathematics is a logical, self-consistent creation of the human mind. It does not need to be tested against messy, approximate reality. Physics, by necessity, is imprecise. How do we know where an object's boundaries are when, with enough magnification, everything dissolves into whirling, colliding atoms and molecules that move in and out of macroscopic objects?

Eventually, things *are* defined; theories are developed and tested against reality. A new theory will have to be built on top of a self-consistent network of known facts and already proven theories that are connected to basic experiments, axioms, and principles. This is the existing **body of science** that, to the best of our current knowledge, does not contain contradictions. This knowledge base is the result of thousands of years of curiosity, passion, determination. It contains knowledge gained in every era and location through history: the Babylonians, the Egyptians, the Greeks, the Chinese, the Arabs, the Europeans, the Russians, the Americans all added data that became 'integrated' along the way.

Integration is a process that constantly compares statements in the knowledge base for consistency and agreement. Whenever a contradiction is found, every effort is made to resolve it. Contradiction in science is like poison in the human body – it has to be expelled for the body to survive. Even if all the experiments performed so far are in agreement with the theory, scientists must be prepared to discard or modify the theory at *any* future time, should contradictory evidence surface in further research.

Once we have a knowledge base, we need the tools to expand it through meticulous observations. The primary tool used by science for learning new things about the physical world is *experimentation*. With experiments, we collect facts, identify the exact nature of what we know and how we know it. Experiments must be repeatable and consistent, publicly demonstrated, and the resultant data freely available to anyone. And, as many teachers and books point out: no measurement is valid without identifying its precision and its margin of error.

An integral part of scientific experiments is called '*reduction*'. We try to determine which parameters play a role in the process we want to study. Once we have an idea about those attributes that affect the process, we have to set up an experimental situation wherein we can change any one parameter and see what happens to the rest.

A good example is studying the relationship between pressure, volume, and temperature of a gas. After many experiments performed by Robert Boyle in 1661 and Joseph Louis Gay-Lussac in 1802, the "Ideal Gas Law" was discovered: the relationship between temperature (T), pressure (P), and volume (V) of a body of gas is such that the magnitude of PV/T remains constant (values determined by the kind of gas we use). This means that:

- If we keep the temperature unchanged and decrease the volume, then the pressure has to increase (V and P are inversely proportional)

- If we keep the pressure unchanged and increase the temperature, then the volume has to increase as well (T and V are directly proportional)

- If we keep the volume unchanged and increase the temperature, then the pressure has to increase as well. (T and P are directly proportional)

While collecting data on a given subject, scientists use both pattern-recognition techniques and *imagination* to find relationships and cause-and-effect links among these facts. This is usually done by trying out different models (*hypotheses, theories*) that could explain the experimental results and not contradict anything we know.

Then the scientist is ready to try to develop a theory that will explain the collected data, based on the existing knowledge base of science. The theory is usually built on an assumption or hypothesis that appears reasonable, in view of all the known facts. Quite often it states this assumption as a hypothetical law or laws. Some of the best-known theories in physics are:

- The Heliocentric Universe theory, suggested first by Aristarchus in ancient Greece, followed by Copernicus in 1514, and then embraced by Kepler, Galileo, and Newton. Galileo got into trouble with the church because he had stated it as a proven theory instead of as a hypothesis.

- Newton's three Laws of Mechanics, coupled with the new mathematical tools he invented, could be used to deduce results in perfect agreement with all known facts at the time.

- Maxwell's Electromagnetic Field Theory offered a set of equations that could be used to calculate all known electromagnetic phenomena. It also suggested experimental results (like radio waves) that were only confirmed later.

- Einstein's Special Theory of Relativity was based on a few philosophical assumptions that explained many unresolved problems and perplexing experimental results. Its predictions have been experimentally confirmed since.

- Einstein's General Theory of Relativity stated a number of equations that explained anomalies in Mercury's orbit, suggested a new view of gravity, and correctly predicted that the sun's gravitational field bends light.

- Heisenberg's Uncertainty Principle states that in the world of atoms, we can never measure everything to an arbitrary degree of precision. For example, the more precise we are in measuring the location of an electron, the less precisely we will be able to measure its momentum - mass times speed - and vice versa.

Actually, Albert Einstein and Niels Bohr spent over 25 years arguing about the Uncertainty Principle. Einstein would dream up thought experiments to prove that the theory was incorrect, and Bohr would prove Einstein wrong every time. Until, one day, at the 1930 Solvay conference, Einstein managed to come up with one example that completely baffled Bohr, who went to bed and spent a sleepless night trying to find a way to prove Einstein wrong, yet one more time. The next morning, they met at breakfast, each with a huge grin on his face. Einstein was convinced that he'd finally defeated Bohr. Bohr, on the other hand, had found Einstein's mistake. Albert forgot to take only one thing into consideration: the effect of his own General Theory of Relativity. There was much merriment around the table that morning!

All the experiments performed since then seem to decide in Bohr's favor. The Uncertainty Principle Theory – as of this moment - is considered to be proven beyond any doubt.

Many other theories have been proposed by physicists during the last 400 years, some fundamental, others minor; some proved correct, many turned out to be wrong. To prove a theory, we need *deductive logic.* By applying mathematical tools, we logically deduce the consequences of the new theory and make *predictions.* We can test these predictions against existing experimental data or perform new experiments to verify the results of the calculation. Once these deductions are tested and their accuracy demonstrated, the theory is considered to be proven.

For example, the "Kinetic Theory of Gases" assumes that gases are made up of atoms or molecules in random motion inside a container. The pressure of the gas is due to the force exerted by the molecules hitting the walls of the container and the temperature is due to the kinetic energy of the molecules. If we apply Newton's Laws to this model, then we can deduce the experimentally obtained "Ideal Gas Law", so the theory is proven within its limits.

Luckily, we don't always have to do a lot of math before realizing that a theory is wrong. A common mistake many young physicists make is taking mathematical deductions too seriously. I don't mean that math can be sloppy and incorrect - far from it. However, as experienced physicists will tell you, it is possible to think in terms of critical and determining variables and see how they stand up in the new theory. Much deductive work can be saved if one can spot a showstopper at the beginning: thinking 'physically' before thinking 'mathematically'.

The theories we make up will be judged by what is called: "*Ockham's razor*". William of Ockham (1280-1349) laid down the rule that "entities must not needlessly be multiplied". Between two theories that fit all observed facts, we accept the theory that requires fewer or simpler assumptions. This does *not* mean that the simplest explanation is always correct. Remember: *all* observed facts must be taken into account!

The famous quote from Albert Einstein is appropriate here: "The most incomprehensible thing about the universe is that it is comprehensible".

Physicists like the elegance and simplicity of nature. They like it so much that they have been pursuing the Holy Grail of Physics for over a century: the "Unified Field Theory". They would like to come up with one theory that explains absolutely everything! A tall order indeed, but there is cause for optimism: as the science of physics progressed over the decades, more and more phenomena that seemed to have nothing in common were proven to be manifestations of the same thing. For example, electricity and magnetism, once considered totally different areas of physics, were unified by James Maxwell in 1873 under the "Electromagnetic Field Theory". Many advances on the 'unification front' have been made since then and physicists still hope that one day they will have one equation that explains the universe!

The Limits of Science

I read a really cool sci-fi story once (Frederik Pohl: *The Tunnel under the World*). A small industrial town was destroyed by an explosion during the night, yet the next morning people woke up, as usual, went to work as usual, and lived their lives as usual. They noticed only one change: commercials and billboards and all forms of advertising became so ubiquitous that there was no escape from it. At the end of the story, they discover that they are nothing but replicas. A ruthless advertising executive had rebuilt the people as minuscule robots and put them into a miniature model of their town on a laboratory bench to use as a testing ground for advertising techniques. Imagine their shock when they made it to the edge of the table and looked down!

I can't help thinking of that story whenever I hear someone being so certain about what the world is and isn't like. We humans like to hold simplistic views. When people think about science, they often hold one of two extreme views:

- Science is omnipotent; can explain everything (if not yet, soon) and everything outside the domain of science (ESP, telepathy, etc.) is pure superstition, to be dismissed out of hand.

- Science is overrated, a useful tool in the material domain, and totally clueless when applied to important spiritual, emotional, and social questions.

I don't subscribe to either of these extremes. I love and respect science and recognize its greatest contribution to human thought in the discipline and methods it gave us for examining any kind of phenomena, be it material or social. On the other hand, I am aware that many human qualities besides reason are important sources of knowledge: intuition, imagination, ethical sense, spiritual insight all play an important role in understanding.

Reason can only give us some probability of being right, as opposed to absolute certainty. It is a very important point, often ignored in debates. Doubt is a useful device (See John Ralston Saul's *The Doubter's Companion*) that keeps us alert at all times, making us examine and re-examine our basic assumptions and the methods we use in pursuing 'truth'.

I find myself in a somewhat 'delicate' situation. This book is primarily about physics, how wonderful it is, what great achievements geniuses made, all implying how great reason is. Then I start writing about the 'limits of reason' and the 'abuse of reason' as if I were contradicting the very theme of this book.

I almost feel like a father who, loving his children very much, is willing to admit that the children are not grown up yet; that they need protection, parental control, and correction. It would be irresponsible to let them run wild, to do whatever they wish because they would hurt themselves and others. As John Ralston Saul wrote in *On Equilibrium*:

> " It [reason] has been used willy-nilly to save lives and
> to take them, to process information efficiently and to
> limit citizens' freedom, to run hospitals and to run death
> camps".

Reason is a marvelous tool if used right, in its own place, within its own limits. Luckily, when reason can't help us, we have other qualities to fall back on.

Our understanding in science is, inevitably, incomplete and sooner or later we are stopped by the boundary of our current scientific experience. This brings us to the essence of what we call science.

What is the essence? Most people identify science as the application of "The Scientific Method" defined by the Random House dictionary as: "a method of research in which a problem is identified, relevant data are gathered, a hypothesis is formulated from these data and the hypothesis is empirically tested." I suspect that the "Scientific Method" has a built-in limitation that will not allow science to go beyond a now fairly visible point.

The question is: if we abandon the 'Scientific Method', can we still talk about science? Or maybe we have come full circle where the task of taking things apart is almost completed and the task of putting it all back together has just begun? In my view, any major progress in science will come from the synthesis and integration of all that we have learned. (See the quote from Richard Feynman below).

In theoretical physics, physicists have been turning to mystical analogies, meditation, arguments of symmetry, harmony, elegance, and beauty, both for inspiration and 'proof' of theories that become more and more 'empirically un-testable.'

Here are a few quotes from leading physicists. mathematicians, philosophers, and other scientists (many of them Nobel laureates) of the last few decades.

Sheldon Glashow (Nobel Laureate, Physics, 1979) "Science is certainly slowing down….for the first time since the Dark Ages we can see how our noble search may end, with faith replacing science once again."

Steven Weinberg (Nobel Laureate, Physics, 1979) "…nor any other earthly accelerator could provide direct confirmation of a final theory; physicists would eventually have to rely on mathematical elegance and consistency as guides….might not reveal the Universe to be meaningful in human terms…all our 'whys' would eventually culminate in a 'because'."

Hans Bethe (Nobel Laureate, Physics, 1967) when asked "Could there ever be another revolution in physics like the one that accompanied quantum mechanics"? – "That's very unlikely".

Richard Feynman (Nobel Laureate, Physics, 1965) "The age in which we live is the age in which we are discovering the fundamental laws of nature, and that day will never come again. *"There will be the interest of the connection of one level of phenomena to another –* phenomena in biology and so on" (my emphasis)

Robert Oppenheimer ("the Father of the Atomic Bomb", Princeton,1965) "The general notions about human understanding… which are illustrated by discoveries in atomic physics are not in the nature of things wholly unfamiliar, wholly unheard of, or new. Even in our own culture, they have a history, and in Buddhist and Hindu thought a more considerable and central place. What we shall find is an exemplification, an encouragement, and refinement of old wisdom".

Werner Heisenberg (Nobel Laureate, Physics, 1932) "The great scientific contribution in theoretical physics that has come from Japan since the last war may be an indication of a certain relationship between philosophical ideas in the tradition of the Far East and the philosophical substance of quantum theory".

Niels Bohr (Nobel Laureate, Physics, 1922) "The search for the ultimate theory of physics might never reach a satisfying conclusion; as physicists sought to penetrate further into nature, they would face questions of increasing complexity and difficulty that would eventually overwhelm them…. For a parallel to the lesson of atomic theory … [we must turn] to those kinds of epistemological problems with which already thinkers like the Buddha and Lao Tzu have been confronted, when trying to harmonize our position as spectators and actors in the great drama of existence"

In the "Scientific Method" section, I have said that the most important prerequisite for the method is intellectual integrity. Follow the data where they lead you, without prejudice. In other words, we need a completely open mind, which, among other things, admits the possibility that *we may never even become aware of some phenomena* in the Universe, and the possibility that understanding some others may be forever beyond our capabilities.

Think about it. We, humans, live on a planet orbiting a star that is just one among billions of other stars in the Milky Way Galaxy, which is just one of billions of galaxies in the known universe, which may be just one of billions of universes. Our recorded civilization goes back a few thousand years on this planet which was formed about 4.5 billion years ago, in a universe that is 13.8 billion years old (give or take a year). And it may be just one of billions of Big Bang - Big Crunch cycles. And we know for a fact, that we are equipped to understand everything?!

Admitting the possibility that we may not be able to understand everything does not require us to give up science or the scientific method. It only means that we keep an open mind and do not reject any reasonable suggestion out of hand, just because it lies outside the currently recognized boundaries of science. In his book, "The Meaning of It All", Richard Feynman gives a delightful account of how he would investigate someone's claim to be a mind reader and be able to prove it. He finishes the story with the most open-minded statement I have ever read from a physicist:

> "To be prejudiced against mind reading a million to one does not mean that you can never be convinced that a man is a mind reader."

Accept it, my friends, we are not gods with infinite power, no matter how much we would like to be. I can't help thinking of a joke, involving patients in a mental institution. I hope no one will be offended:

There is a tall pole in the middle of the exercise yard. The patients affix a board to the top of the pole, with a note on it, and climb the pole every day, one by one, read the note, nod, then descend.

The staff is burning with curiosity. Finally, one night, after the patients retire, one of the doctors climbs up the pole, reads the note, nods, and descends. "What does it say"? asks the other doctor. "It says: 'this is the end of the pole, don't try to climb any farther." the first doctor replies. They both nod and go home.

Beyond Science

Physics often humbles me -- I am awed by the beautiful rationality of what I see. Where do these "Laws of Nature" come from? Why does inanimate matter follow them with unerring precision?

Take one example: The law of conservation of angular momentum. It basically says that as long as there is no external torque acting on a system, its angular momentum (rotational inertia) must stay the same.

Look at the planet Earth, or even Jupiter, whose mass is enormous compared to Earth's – almost a sun itself – obeying slavishly the 'commandment': "Your angular momentum shall remain the same". This enormous mass speeds up as it gets closer to the sun and slows down as it gets farther away, in a precise, mathematically consistent manner, as discovered by Kepler in the 17th Century. If you asked today's physicist why it is so, he will tell you: "it is a fundamental law of nature, as predicted by Newton's second and third law". If you ask him why this is so, he will tell you, he has no idea, his job is simply to describe nature as he sees it.

Fair enough.

In practice, the law can often produce surprising results.

Look at a figure-skater. She starts a pirouette by giving herself a small spin on the ice, with her arms stretched out. Once she has a low-speed rotation, she gradually pulls her arms in, lifting them slowly to a vertical position, until her arms are as close to the axis of rotation as possible. Observe how her spin speeds up as she does that. Her body, without any further effort on her part, obeys the law: "Your angular momentum shall remain the same" - the law of conservation of momentum must be obeyed. Just imagine how a little girl must feel when she first learns this neat trick: without any effort, just by raising her arms, a sudden force grabs her and starts spinning her body in a dizzying pirouette!

Here is another experiment anyone can perform. Stand on an old-fashioned piano stool, holding a bicycle wheel, each end of its axle between the thumb and forefinger of one of your hands. Then give the wheel a rotation by pushing on the spokes with the rest of your fingers. As the wheel turns, slowly tilt the axle to the vertical position. The piano chair will suddenly start turning under you, in the direction *opposite* to the spin of the bicycle wheel.

As if the piano chair 'thought': "I must obey the Law. Since there is a sudden increase of angular momentum, I must compensate by turning in the opposite direction." When I showed this experiment to my son, he was utterly amazed, because he had expected the piano chair to start turning in the *same* direction as the bicycle wheel.

Inanimate matter acts in a predictable way, according to certain principles that are marvelously logical and marvelously consistent.

Physics gives us a lot more than formulas, rules of clear thinking, and methods for discovering the truth. Physics is breathtaking in its scope and depth.

Let me tell you what physics means to me. It is a door to a universe full of secrets and miracles and mystery. It is Alice's Wonderland, where everything is different from what you expect. Space and time are one and curved; simultaneity is an illusion; forever can last only a second, and a minute may last forever. Cause and effect may change roles and, if you didn't pay attention to the sound of the falling tree, you may have made it 'never happened'.

Physics is a window to look at the birth - and death - of the universe; it allows you to read messages from billions of years in the past, before the atoms making up the planet you stand on came into existence. Physics is the rapidly increasing heaviness that paralyzes your body as you approach the speed of light so you can move only as in a dream - but only to the observers, you are speeding away from, while you think you are going around in your normal way. It is a world where matter can be created from nothing but energy and then turned back into radiating emptiness again; where particles can annihilate each other, winking out as they collide; where either of 'twin' particles, moving in opposite directions at the speed of light, still knows and reacts to what happens to the other; and where the entire universe may be nothing more than fluctuating vacuum, delineated by the Big Bang and the Big Crunch.

How can anything compete with miracles of this kind?

Physicists are acutely aware of these implications, as Paul Davis's *The Mind of God*; Fritjof Capra's *The Tao of Physics,* or Gary Zukav's *The Dancing Wu Li Masters* tell you. Even physicists less bent on spiritual interpretations are subject to the awe of unknown forces. Scientists know how mysterious and powerful nature is. Most of them don't need gurus to attain faith. They don't wander aimlessly through life, trying to fill time with something that does not feel artificial and pointless. They have close contact with something infinitely larger than themselves; they can feel one with the living, breathing, sustaining reality that surrounds us all. Some of them are religious and call it God; others just call it Nature – both agree: *it* is great and awe-inspiring, mysterious yet dependable, clear and logical and, above all, beautiful.

Logical thinking

Most people would agree that an essential ingredient of reason is logic. How many of us could clearly define logic and summarize its rules? There are only a few rules but many reasons to use them all the time in every area of life.

Aristotle laid down these rules circa 300BC. In 1637, Rene Descartes published his own version, titled *Discourse on the Method of Rightly Conducting one's Reason and Seeking Truth in the Sciences*. Bertrand Russell formalized and connected logic to mathematics in the 1940s. They did not invent logic, merely 'documented' it: analyzed their own thinking and described what made sense to them.

The basic rules are:

The Law of Identity: Everything is what it is - and cannot, at the same time, be something else.

The Law of Excluded Middle: Every statement, exactly as it is stated, is either true or false.

The Law of Non-Contradiction: No statement may be simultaneously true and false.

All three of these laws state that contradictions cannot exist in our Universe. Our hardwired brains just cannot conceive of a situation where any one of these rules could be violated.

The two main concepts related to logic are Inference and Argument.

Inference is our inner psychological reasoning process when we, based on available data, form an opinion or arrive at a conclusion.

Argument is the externalizing of inference: in an argument, we explicitly state both the evidence (the statements making up the premises of the argument) and the conclusion (the statement that the evidence is supposed to prove). An argument is a group of statements all but one of which are designated premises, and one of which is designated the conclusion. An example of an argument is:

Premise 1: All persons are mortal
Premise 2: Bob is a person
Conclusion: Bob is mortal

Another example of an argument:

Premise 1: I am six feet tall
Premise 2: The Moon is orbiting the Earth
Conclusion: I like to eat ice cream

In the first argument two TRUE and related premises result in a TRUE and *valid* conclusion. In the second argument, even if both premises and the conclusion are true, it is an *invalid* argument, because the premises have nothing to do with the conclusion. Logic deals with the relationship between the premises and the conclusion, not with the truth of the premises.

There are two different types of arguments: Deductive and Inductive. Each has logically correct and incorrect forms. Correct *Deductive Arguments* satisfy the following criteria:

- If all the premises are true, then the conclusion must be true

- All of the information in the conclusion must be contained or implied by the premises. Nothing new is allowed in the conclusion.

Correct *Inductive Arguments* satisfy the following criteria:

- If all the premises are true, then the conclusion is probably true

- The conclusion contains new information not even implied by the arguments.

A correct deductive argument is called "*valid*".
We classify arguments in terms of their *form*. The most commonly used form is the "Conditional Argument": one or more conditional statements and true/false statements about the conditions. The form of an argument can be valid or invalid.

A *valid form* is one where it is impossible for any argument of that form to have true premises and a false conclusion. Any argument that has a valid form is a valid argument.

An *invalid form* of argument is called a *"fallacy"*. In order to prove that an argument is fallacious, it is sufficient to find a counterexample: an argument of the same form with true premises but an obviously false conclusion.

We must be aware of typical fallacious argument forms in order to keep our own thinking clear and logical and to avoid believing misguided or malicious arguments and accepting an invalid conclusion. Let's look at some examples:

Conditional argument valid forms:

	Form	Example
Premise 1	If p then q.	If I fail the exam I am toast
Premise 2	p	I failed the exam
Conclusion	q	I am toast

	Form	Example
Premise 1	If p then q.	If I fail the exam I am toast
Premise 2	Not q	I am not toast
Conclusion	Not p	I did not fail the exam

	Form	Example
Premise 1	If p then not q.	If I pass the exam I am not toast
Premise 2	q	I am toast
Conclusion	Not p	I did not pass the exam

Here are a few fallacious forms of Conditional arguments:

	Form	Example
Premise 1	If p then q.	If I fail the exam I am toast
Premise 2	q	I am toast
Conclusion	p	Therefore I did fail the` exam

This is an invalid argument because I could be toast for any other reasons than failing the exam. The premises do not prove the conclusion.

	Form	Example
Premise 1	If p then q.	If I am in California, then I am in the US
Premise 2	Not p	I am not in California
Conclusion	Not q	Therefore I am not in the US

This is an invalid argument because I could be in the US even if I am not in California.

A typical example for the last logical fallacy is a politician telling the crowd: "If you want to pay higher taxes, then you should vote for my opponent. But I know that you don't want to pay higher taxes, therefore you should vote for me". It sounds logical, doesn't it? Yet, it is totally fallacious. Beware!!!

There are many typical mistakes people make in logical deductions, sometimes innocently, but often deliberately, with intent to mislead. We have to be aware of these mistakes both in our own thinking and in arguments presented to us. My favorite examples for illogical thinking these days involve political and military leadership:

"Canada won't join missile defense plan" Last Updated Thu, 24 Feb 2005 21:32:18 EST <u>CBC News</u> OTTAWA

"Canada has said no to the U.S. missile defense program, Prime Minister Paul Martin announced Thursday. The outgoing U.S. ambassador to Canada reacted swiftly, saying the decision to defend North America now rests with Washington. "We simply cannot understand why Canada would in effect give up its sovereignty – its seat at the table – to decide what to do about a missile that might be coming towards Canada," said Paul Cellucci. "

The 'illogic' in this example is the following:

	Form	Example
Premise 1	If p then q.	If you can join a decision-making process then you are a sovereign country (TRUE by definition of the word sovereignty)
Premise 2	Not p	Canada did not join the decision-making process (TRUE)
Conclusion	Not q	Therefore Canada gave up her sovereignty (FALSE)

In mathematics, we say that 'p' is a sufficient but not a necessary condition. The form of this logical argument is invalid. Canada could have said yes' but decided not to. Therefore, Canada is a sovereign country. Again, the premises don't support the conclusion.

Fallacies can be presented in two basically different ways:

- One or more of the premises are false
- The form of the argument is logically invalid

Therefore, when we hear an argument, it is crucial to examine both the correctness of the premises and the validity of the logical form.

It is important to insist on sensible logic and refuse to be blinded by an argument that is formed as if it were logical while violating the rules of logic. As they say: "If you live in a dangerous neighborhood, you had better learn as many self-defense techniques as you can."

Science and Society

I often hear people argue about their country. I have heard educated, intelligent people try to explain it on television, in books, in speeches. What bothers me is the almost total lack of reference points. The body of explanation is floating in the air, without an anchor in reality. We have no starting point on which we agree.

If I ask anyone in Canada to define the basic principles by which our society is organized, I get different answers. Some say it is a democracy, some say it is capitalistic; most would agree that it is a nice place to live. (Canadians are proud to be 'nice people'). If I ask about the problems we have, I really get an earful: too much taxation, too much government, too few public services, too soft on crime, too much corruption, too much poverty, etc., etc.

And when I ask, "too much, compared to what?" - no answer. How do we define what is normal, what is acceptable? According to what principle; by what yardstick? Nobody knows. We just don't like some part of our social environment for personal reasons and we call it too 'something'.

Take taxation. Everyone agrees that we pay too high a proportion of our income. What would be the right proportion? Why pay taxes at all? And if we do, what is the appropriate amount? Should it be the same for everyone or should it be progressive? If yes, how progressive? Why? What determines fairness? How do we calculate it? Everyone who ever filled out a tax return knows that tax laws are insanely complicated. Who made them, based on what principle, what criteria?

Almost nobody, at least not in public media, discusses these questions. We just express emotional and personal opinions and expect others to agree with our unstated assumptions. Sometimes even we ourselves do not know what assumptions.

Our opponents are no better off, so arguments seldom go anywhere. We keep shouting each other down, interrupting each other's statements -- nobody convinces anybody about anything; the argument is doomed from the start. Quite often the purpose is to score points. We treat the discussion as a contest, instead of an attempt to find a solution and thus let everybody win. This attitude, of course, is consistent with the aggressive genes in our species that want to fight, rather than cooperate, for survival.

If we tried to build science and industry by this method, we would still be in the caves. It just doesn't work. It can't. The scientific method, which was so successful over the centuries in technology, is not limited to science: it is a general problem-solving method that could and should be applied to all our problems.

We need a common starting point. If we go from there, step by step, making sure we agree on each step, then either we arrive at the same conclusion or a point of disagreement. Work on that point, until we reach a compromise, and then resume our discussion, knowing that we are still together, solving our problem.

In the case of taxation, we would have to ask some basic questions first, before going into details or percentages:

What are the essence, goals, and priorities of society?
What are the basically different options for organizing people?
What are the advantages and disadvantages of these options?
Which of the options do we choose?
What is the optimum way to implement this option?

If we answered these questions, the rest would be easy. Basic facts, simple logic, and some arithmetic would provide the answers.

The general public assumes that the scientific method is designed for, and restricted to, science. Nothing could be farther from the truth.

Take the judicial system, for example. The body of laws ought to read like a scientific document. All the terms must be clearly defined, all the laws clearly stated, covering every probable scenario, every possible exception. No contradiction is allowed in the document and if one is demonstrated, it needs to be revised to remove the contradiction. Of course, no law book is perfect, just as no encyclopedia of science is flawless. But the intent is there and with the right attitude, things can be improved all the time.

In criminal trials, both the defender and the prosecutor have to use precise logic to draw their conclusions (even though each tries to cheat as much as he thinks he can get away with) and the evidence they present has to be "beyond a reasonable doubt".

It is a sad state of affairs that our politicians can get away with undefined concepts, gross errors of facts, blatantly illogical arguments, glaring contradictions, and transparent emotional manipulation. Just compare the public 'debate' that led to attacking Iraq in 2003, with the process I described above and ask yourself the following questions:

- Were all the terms used clearly defined
- Were all the relevant data considered?
- Were all the statements offered consistent with one another?
- Were all the presumed facts clearly demonstrated?
- Were the conclusions reached by meticulous logic?
- Has the need for, and goals of, action been clearly identified?
- Had every alternative action been considered?
- Were the leaders ready to admit error when contradiction was found?
- Was the course of action changed according to new evidence?
- Was there an attitude of honesty, integrity, openness, objectivity?

Somehow, I think that it would help if the citizenry were better educated in science and logic. This is part of the reason I decided to write this book.

<u>What is, and how to find, Truth?</u>

Truth is an observation or a plausible theory that has not [yet] been contradicted by the accumulated knowledge available to us. The very instant an irrefutable contradiction is demonstrated; the theory becomes false and needs to be modified or discarded.

So, Newton's equation: "force equals mass times acceleration, where mass is an unchangeable constant attribute of material bodies", was a truth until Einstein, by his special theory of relativity, suggested experiments that produced results contradicting Newton. At that point, Newton's 'truth' became 'false' and required modification.

Truth can apply to physical observation, directly with our senses or indirectly with instruments, or to a theory (which is usually a statement about cause-and-effect relationships). Physical observations have to be repeatable and consistent at ever-increasing accuracy, to be considered true; theories have to produce predictions verifiable by observation.

The only intolerable state is a contradiction.

Life is not different from science - only more complicated. Basic principles still apply. A while ago, telling my students how to find the truth, I used the example of Hercule Poirot. Imagine that you are a detective. A murder has been committed and you need to find out the truth. You question suspects and witnesses. Some tell the truth, some lie; you have no idea which. You collect all the evidence, all the statements from those interviewed, and build a model in your head. At that point "the little grey cells" ought to do their job.

You arrange the known facts and statements in such a way as to have a minimum number of contradictions in the model. You discard those that cannot be used without contradicting most of the other facts. Then you try to come up with a theory that agrees with the largest number of 'facts' and is supported by your personal experience, the experience of those you trust, and discard as 'untrue', all the rest. Then you think you know what the truth is. You draw logical conclusions and test these in real life. If they check out, you can be reasonably sure. You will never be absolutely sure -- truth is still only those theories that have not yet been contradicted.

Applying this method to our understanding of the human condition is no different. The keys to reasonable confidence in knowing the truth are to

- have extensive personal life experience (needs observation)
- learn as many facts as you can (needs a *lot* of reading)
- keep all of these facts in mind (needs a good memory)
- try to form a theoretical model (needs pattern recognition ability)
- be *completely* open-minded (needs intellectual integrity)

Then you can be reasonably sure, in the relative sense. In the absolute sense, you can be only 50% sure. Either you are right, or you are wrong. However, we can only do the best we can do. For all practical purposes, I can call it truth: 'my truth'. And it stays true until someone proves it false.

Unfortunately, when the topic of *truth* comes up, most people assume we are talking about metaphysics or quantum theory and the conversation becomes highly philosophical and highly impractical. To consider Heisenberg for my modest purpose of learning the truth about the reason we are unhappy and the changes we have to make to set it right is like considering Einstein to plan a car trip to Florida.

We also have to deal with another misconception of our age: it basically claims that no one has a *by definition* superior value system or opinion; that my opinion is just as good as someone else's. While I understand the necessity of devaluing arguments based solely on authority, I find that we managed to swing to the opposite extreme (as usual) and tried to be completely egalitarian when evaluating competence and ability.

There *are* more and less intelligent people, more and less educated, experienced, knowledgeable, wise, honest, etc. people in the world. It is still true that *some* younger people could learn something from *some* of their elders, *some* experts still know more about their fields than *some* laymen.

Convictions are not entirely relative with identical weighing factors. Before I evaluate it, I like to know how a person arrived at his opinion, what it is based on, how many factors were considered, how much factual knowledge it incorporated.

Science and Religion

We have seen how science attempts to acquire knowledge. It is a long-established method that has produced tangible results: our lives depend on it every day.

When we talk about religion, first we have to ask: which religion? There have been hundreds in human history and countless people believed in each of them, convinced that their religion was the only true one and all the others were deluded. Just Google "World Religions" for a sample of dozens still practiced today by millions to billions.

I used to participate in internet forums for the discussion of scientific and philosophical ideas. On one of these forums I posted the following question:

"Would you have imagined a god if you had never heard of the concept?"

Suppose for argument's sake that you grew up in a world where nobody ever talked about gods or supernatural of any kind. Suppose you had a totally secular education: you learned about nature, physics, scientific facts, technology, productive skills, social organization, project management, etc. No priests, no churches, no bibles, no superstition, no Santa Clauses, no tooth fairies - nothing but observable reality. Would you have ever thought of anything outside this? What, if anything, would have made you think that there might be something outside of your experience?

The point I was trying to make is that ALL of our 'knowledge' concerning religious assertions were handed down to us by our cultures. None of us discovered it from personal experience. This question made many forum members think hard, asking themselves the same question: "What do I know from first-hand experience and what have I accepted from others, without really examining how they acquired that 'knowledge'?"

You might be tempted to say that the same is true for science: after all, we learn it from textbooks written by others. However, there is a difference. We can find out how the authors made their discoveries, based on what experiments, and how they reached their conclusions. Interested amateurs can reproduce the simpler experiments themselves, at least in the domain of Classical Physics. You need not take anything on faith.

Obviously, there are historical reasons why religions were invented in the first place, thousands of years ago. Otherwise, they would not exist today. However, religions were established before we had proper science as an alternative and superior way to explain the universe.

The reason science has not replaced religion in so many minds is that people often lie, are often deluded and, the saddest fact of all, they often use psychological manipulation to achieve their aims: wealth or power over other people. Religions have been used for both over the millennia. Many bloody wars were fought using religion as an excuse.

In view of this, how much should we trust religious assertions, handed down to us over history? Wouldn't it be safer to rely on our own observations and our own minds? Scientific thinking offers exactly that.

I was once asked whether I 'believed in' electrons. My answer was: I don't need to believe in electrons, because I have personally conducted experiments that proved to me that material particles with a definite mass, charge, and spin exist, even if I can't see them. I don't believe – I know.

The other argument I often hear is based on a lack of imagination. It goes like this: "How can you imagine that a world as complex and as perfectly interacting as ours, has evolved by chance? There had to be a creator".

And, of course, this reply begs the question. If the world was created by a creator, then the creator had to be at least as complex as its creation. Then, using the same argument, the creator had to have a creator, so who created the creator?

The usual answer is: the creator has always existed, it was not created. Then, the question is: if we can assume that something complex and powerful always existed, then why can't we just assume that the universe has always existed, without a creator? Whichever way we look at religion, we either run into contradictions or find ourselves inventing arbitrary and totally unnecessary concepts.

Science saves you from all these problems: it is simple, logical, available to everyone who wants to find out. You don't have to take it on faith.

Bottom line: am I an atheist? If the word 'atheist' means that I am absolutely certain, beyond even a shadow of a doubt that there is no such thing as a 'god', then I am not an atheist. No self-respecting scientist can be 100% certain of anything in the universe. Only probabilities exist in science and I admit, for lack of evidence to the contrary, that I assign an extremely low probability to the idea of a creator.

However, nothing is *proven* one way or another. Yes, the universe could have been created by a god or any number of gods. Life and evolution could have been started on Earth by an alien culture of superhuman power and we would not know anything about it.

However, all the established religions with which I am familiar are so obviously man-made that I find it difficult to believe that anyone could take any of them seriously. Charles de Secondat, Baron de Montesquieu said: "If triangles had a god, he would have three sides".

Chapter II – Early History

We have to start with ancient Greece because, in a way, it all started there 2700 years ago.

The Greek experiment

Human history is mostly depressing stuff. Greed, vanity, envy are the main forces motivating the powerful -- duplicity, ignorance, violence are their favorite tools. But every now and then, for brief periods, the trend is reversed, and we see curiosity, creativity, honesty, integrity, and fairness in the ascendant. The consequent major advances in all areas of human endeavor astonish us all. These are the ages when intelligent thinking is admired when pride in human ability and optimism for the future prevail.

The first of these periods in recorded western history took place in ancient Greece between circa 600BC and 100AD. It followed, significantly, perhaps the most dramatic example of an act of intelligent problem-solving in the social arena, as described by Will Durant: in *The life of Greece - The story of civilization*. When faced with popular revolt due to the citizens having been mercilessly exploited by the moneyed classes, Solon passed a law abolishing all debts owed either to private persons or to the state. Imagine someone trying to suggest a solution like that today!

The heroic age of Greece lasted from Solon's saving of Athenian society in 594 BC, until about 125AD when Ptolemy carved the heavens into epicycles for the next thousand years. The period was not one of uninterrupted achievements. Seeds of its downfall were planted all through the era: by Pythagoras's mysticism and secretiveness; by Plato's contempt for curiosity about, and investigation of, the material world; by Aristotle's authoritarian dogma regarding the perfect circles and the crystalline spheres of the heavens. But, while it lasted, the ancient Greek miracle was spectacular.

- Pythagoras (580-500BC) elevated Mathematics to its proper role as the language best suited to describe nature and reality.

- Democritus (460-370BC) speculated about atoms as the building blocks of matter.

- Herakleides (375-370BC) suggested that at least Mercury and Venus orbit the sun.

- Eudoxus (408-355BC) created the first star map and invented longitude and latitude to pinpoint the locations of stars.

- In 300BC the famous *Museum and Library of Alexandria* was founded and became the focal point of science and scientific research.

- Euclid (325-275BC) published his *Elements of Geometry*, which is still valid and taught today.

- Aristarchus (310-230BC) calculated the size of the sun and the moon and suggested the first model of a heliocentric (the sun's Greek name is Helios) universe.

- Archimedes (287-212BC) discovered the precise rules concerning weight loss in water, invented and described levers, pulleys, and the water pump, described center of gravity, and calculated the value of pi, using methods reminiscent of calculus invented by Newton and Leibnitz almost 2000 years later.

- Eratosthenes (276-196BC) calculated the size of the Earth, the sun, and the moon.

- Hipparchus (146-127BC) calculated the distance of the moon, created a much-improved star map, and invented trigonometry.

The decline began after the Roman conquest because Rome, while it excelled in engineering, did not support pure science. When a scientist's or mathematician's skills were required, Roman rulers hired a Greek. For example, the 'Julian' calendar was devised by astronomer Sosigenes of Alexandria in 46BC.

Aristotle (384 - 322BC)

No philosophers have had such a strong and enduring influence on Western thought and science as Plato and Aristotle. They influenced European thought for a little over 2000 years, in alternating Platonic and Aristotelian waves.

Plato's effect was devastating - he despised science and empirical observations – for him only ideals were real.

Aristotle, on the other hand, approved of science. He not only approved of it but turned it into an edifice, a finished piece of architecture, forever stamped with the brand of authority and dogma.

Aristotle on Physics is boring and pedantic. It boggles the mind how someone of his genius could talk so much about something he knew so little. He never observed, never experimented, never even looked up from his parchments – he just speculated idly about what the world might look like. Even the proverbial six blind men at least *touched* the elephant once!

Aristotle's physical universe is a static, artificial creation of concentric crystalline spheres around Earth, like the layers of an onion, housing the planets, the sun, and the moon. On the outermost layer are the 'fixed stars'; beyond that is the 'Prime Mover', who rules over and controls everything.

In the innermost sphere is the moon, and at its center, the Earth. This is the only place in the universe where change is allowed. In the sub-lunar region, all matter is made up of four elements: earth, water, air, and fire, constantly transformed from one to another. This is the process solely responsible for the change. Beyond the moon's sphere, everything is made up of a pure fifth element: aether, which is eternal and unchangeable.

All movement in the heavens is circular because the circle and the sphere are the only perfect forms. Nothing moves unless something moves it, and this mover, in turn, can also be moved, until we reach the end of the chain: the unmovable 'First Mover'. As soon as a mover stops moving, the moved object instantly comes to a complete rest.

The weight of objects on Earth, he attributed to the heavy object's natural 'desire' to move towards the center of the Earth. The rising of smoke was due to a light object's natural 'desire' to move toward the heavens. Heavy objects, having greater desire, fall faster than light ones. Obviously, Aristotle never thought of experimenting to determine how nature really behaves. In the area of physics, he relied on speculation alone. His accomplishments in Biology are much more impressive.

Aristotle founded his academy, called the Lyceum, in 335 BC. He and his disciples published about 150 books, 50 of which survived the sacking of Athens, and were found in a pit in Asia Minor by another Roman army circa 80BC. He was highly valued by the Arabs, in whose keeping his books survived European upheavals until they were rediscovered during the Renaissance by Christian Europe and translated into Latin.

It is important to understand Aristotle's physics because of its powerful influence on scientists, even in Galileo's time. Part of Galileo's problems came from the influential "Aristotelian Brigade" in academic positions throughout Europe. Arguably, Copernicus, Kepler, and Galileo had more to fear from this group than from the Catholic Church itself.

Ptolemy (87-150AD)

Ptolemy was another very influential Greek astronomer whose ideas dominated that science until Copernicus who, in an attempt to improve upon it, started the process of its dismantling and final destruction in the 16th century.

Building on works of other Greeks before him, such as Hipparchus, Ptolemy developed an Aristotelian universe of concentric circles (instead of spheres) centered around Earth. He published his system in a detailed and impressive book that Arab scholars later called "Almagest" (The Greatest).

Ptolemy's circles contained the moon, Mercury, Venus, the sun, Jupiter, and Saturn. Each heavenly body was assigned several secondary circles, called epicycles, to orbit on, in a Ferris wheel configuration. The wheel itself orbited along the main circle, called the deferent, with Earth in its absolute center. This elaboration was necessary to account for the observed irregular movements of the planets. Only uniform, circular motions were allowed, as he stated quite explicitly:

"We believe that the object which the astronomer must strive to achieve is this: to demonstrate that all the phenomena in the sky are produced by uniform and circular motions…Having set ourselves the task to prove that the apparent irregularities of the five Planets, the sun and the moon can all be presented by means of uniform circular motions because *only such motions are appropriate to their divine nature*…We are entitled to regard the accomplishment of this task as the ultimate aim of mathematical science, based on Philosophy".

This system, while very complicated - and having no connection to reality - was used for over a thousand years to predict the positions of the planets.

Thus, the process started by Aristotle and 'mathematicised' by Ptolemy, set the 1000+ years of obsession with a perfectly uniform circular motion in stone. No one was willing to question it until Copernicus came upon the scene.

1000 years of darkness

Why the Greek experiment came to an end has been debated over the centuries. Carl Sagan blames the institution of slavery for its decline. Arthur Koestler suggests Plato's influence as the main culprit. Roman conquest and Rome's indifference to pure science were mentioned by Will Durant. The spread of Christianity and the need for suppressing independent thought is often cited as the main reason.

St. Augustine (354 - 430) for example, wrote in *City of God*, published in 413, that humanity is a heap of depravity, in a state of original sin, and salvation could be found only in contempt for all science and curiosity, as well as contempt for all forms of pleasure (food, music, beauty, etc...). Historians' descriptions of the symptoms of decline in the Middle Ages sound frighteningly familiar today:

- the gradual disappearance of curiosity
- disuse of basic rules of logic
- uncritical acceptance of unsupported claims by Authority
- appeal to emotions: fear, greed, envy, hate, intolerance
- mental laziness and suspension of intellectual effort
- jeering, and worse, aimed at dissenters
- crusades and wars against heresy
- spying and persecution, torture and execution, by rulers

The wave of discoveries - both in theory and in practical inventions - came to a halt, followed by a gradual mental decline, in a regression to the mysticism and superstition of the pre-Greek era of Egypt and Babylon.

In Isaac Asimov's *Chronology of Science and Discovery,* there are 43 entries for the period 585 BC to 140 AD (mostly Greek) but only 23 entries for the period from 180 AD to 1200 AD (mostly Arab and Chinese). A sad report card, indeed!

The thaw after the 1000 years long dark ages began, tentatively, in 999AD when the scholarly Pope Sylvester II (Gerbert d'Aurillac, 946-1003) reintroduced Europe to mathematics, astronomy, and the abacus. He was first to use the decimal system with Arabic instead of Roman numerals, a wooden sphere to represent Earth. This was a major innovation, considering that in 550 AD, Cosmas, in his still widely studied *Christian Topography,* depicted the Earth as flat, rectangular, twice as long as it is wide – having the shape of the Holy Tabernacle.

The revival started speeding up when some ancient Greek documents were rediscovered: in 1120 AD, Euclid was 'found' again; in 1175, Ptolemy's book reappeared. Then gradually, ancient texts by Aristarchus, Hipparchus, and Democritus became available. As more Catholic scholars spent time in Constantinople and Cordoba, philosophical inquiry, along with technological advances, spread through Europe during the 12th century. Aristotle once more came into fashion. Albertus Magnus (c.1200–1280, eldest son of the Count of Bollstädt) interpreted, systematized, and brought into concordance with Christian doctrine, those works of Aristotle that had been translated into Latin.

Two major influences elevated science and rational thought to a place of respect. A generation of scholars, among whom Roger Bacon (1214-1294) and Thomas Aquinas (1225-1274) stand out, they praised reason, observation, science, and mathematics as the main sources of knowledge. Progress got a big boost in 1454 with the invention of the Gutenberg printing press. It produced a torrent of scientific and philosophical manuscripts, which became available to thousands and replaced the hand-written parchments that had been read only by a privileged few.

Suddenly science, mathematics, technology was everywhere.

- Leonardo Da Vinci (1452-1519) invented almost everything: military machines (submarines, tanks, planes), parachutes, and elevators; correctly described inertia and the acceleration of falling bodies; set the Earth on a spinning, moving orbit; recognized that the moon reflects sunlight; devised code for his manuscripts.

- Nicolas Copernicus (1473-1543) revived Aristarchus's idea of a sun-centered universe and created the first mathematically based heliocentric astronomy.

- Tycho Brahe (1546-1601) revolutionized the science of astronomical measurement and observation.

- Francis Bacon (1561-1626) took up the subject of logic where Aristotle left it almost 2000 years earlier and developed a modern method of reasoning and science.

- Galileo Galilei (1564-1642) carried it further with his precise formulation of the "Scientific Method" and applied it to the study of moving objects on Earth.

- Johannes Kepler (1571-1630) unified Brahe's observational data with his own mathematical skills and produced the first exact laws of planetary motion.

- Rene Descartes (1596-1650) married algebra to geometry, thus preparing the ground for the invention of calculus.

- Pierre Fermat (1601-1680) taunted the great mathematicians of Europe with his virtuoso solutions to the most difficult problems in number theory.

- Evangelista Torricelli (1608-1647) experimented with vacuum and air pressure, eventually leading to the barometer and the steam engine.

- Blaise Pascal (1623-1662) discovered the basic principle of hydrostatics, founded (with Fermat) the mathematics of probability and confirmed Torricelli's theory of atmospheric pressure.

- Robert Boyle (1627-1691) built an air pump to produce a harder vacuum, proved that in vacuum all objects fall at the same rate, confirming Galileo's theory, experimented with gases and discovered a law relating pressure to volume, became a convinced atomist, founded modern chemistry and laid down the now 'sacred' rule for a precise description of experiments.

•   Christiaan Huygens (1629-1695) invented better telescopes and discovered Saturn's rings and its moon Titan; he made the first measurements of distances to stars, invented instruments to measure special angles and the first pendulum clock; he developed the wave theory of light and suggested the conservation of both momentum and kinetic energy.

•   Robert Hooke (1635-1703) experimented with springs, worked on the wave theory of light, theorized about gravitational forces, performed experiments and described observations in biology and discovered cells in the porous structure of cork.

•   And, finally, Isaac Newton (1642-1727) burst on the scene with his incredible mind and firmly established science and the scientific method in the form familiar to us today.

Newton's greatest achievement is what we call today Newtonian Mechanics, the science of moving material bodies, from planets, comets, and the moon to oscillating violin strings and suspension bridges.

Newton's System is so powerful and such a fundamental theory that he was the only physicist besides James Maxwell and Albert Einstein to be included in Richard Feynman's table of "Classical Physics" in his "Lectures on Physics", Volume II Table 18-1.

Newton's and Maxwell's theories were separated by a gap of 186 years. Nothing as fundamental as Newton's theory was produced in all that time. This shows that giants like Newton, Maxwell, and Einstein are rare treasures in human history. The least we owe them is to understand what they tried to teach us.

Newton greatly impressed his contemporaries with his explanation of the solar system, the working of which had puzzled men for millennia. His explanation is as valid today as it was over three hundred years ago.

Physics students these days don't pay much attention to astronomy. A pity, actually, because it is quite a shock to *fully* realize we *are* a speck of dust in an infinite Universe. When I travel by air and look out the window from 40,000 feet up, I feel awed by the detachment. I almost taste the infinity of the void around me. Above, I can actually see the blackness of space; below, I can see the horizon slightly curved - the contour of the planet under me. Modern physicists today are preoccupied with the mysteries of relativity quantum physics, unified field theories. They don't wonder much about the stars unless their special branch is cosmology.

Understanding the birth of physics requires us to realize that there were no relativity or quantum theories in Newton's time. Even electricity was represented only by the occasional lightning that ripped across the heavens. Human perceptions were more basic: sound, heat, light, movement – and the sky.

The sky, with its moving lights, held endless fascination for the ancient philosophers. It was full of significance; of omens for the future, promises and threats of unimaginable proportions; a place where gods lived… and for a very few, a puzzle to solve. Physics began with Astronomy. It was the movement of the planets, those strange wandering stars that presented an irresistible challenge.

Not fully understanding what transpired before Isaac Newton can give the student an incorrect impression that all scholars before 'real science' were confused, superstitious, or at least insignificant. Nothing could be further from the truth. When Newton started thinking about the motion of planets and cannon balls and apples (*if* he did think about apples), many of the factors had already been worked out by other scientists, even though there was a profound confusion regarding how the pieces fit together.

The most influential of the pre-Newtonian scientists were Copernicus, Brahe, Kepler, Galileo, and Descartes.

In order to understand any great achievement, it is necessary to look at the world as it was before. What made the achievement possible? What questions did it answer, what problems did it solve? What was made possible by it? What is its essence and deepest significance? I will try to answer these questions on the following pages.

Nicolas Copernicus (1473-1543)

Copernicus was originally motivated by a desire to improve on the Ptolemaic system because Ptolemy had allowed for some movement at not-quite-uniform speed in his scheme and Copernicus found this imperfection distasteful.

"Having become aware of these defects, I often considered whether there could perhaps be found a more reasonable arrangement of circles … in which everything would move uniformly about its proper center, as the rule of absolute motion requires".

Isn't it ironic that the name most closely associated with the start of modern Physics, the 'father' of the heliocentric universe, the person almost synonymous with 'vision' and 'courage' is, after all, a fallible human being who wanted nothing more than to correct an aesthetic blemish on the Ptolemaic system of astronomy? As he set out to do that, he was totally unaware that he put some very powerful wheels in motion that would eventually destroy what he wanted to perfect. Science, as indeed human history, follows weird and wonderfully winding paths toward truth.

So, Copernicus started considering an ancient Greek idea of the sun at the center of the universe. He summarizes his assumptions in seven axioms that he thought would address "this very difficult and almost insoluble problem"

1. the heavenly bodies do not all move round the same center
2. the Earth is not the center of the universe, but only of the moon's orbit and of terrestrial gravity
3. the sun is the center of the planetary system and therefore of the universe
4. compared to the distance of the fixed stars, the Earth's distance from the sun is negligibly small
5. the apparent daily revolutions of the firmament is due to the Earth's rotation on its own axis

6. the apparent annual motion of the sun is due to the fact that the Earth, like the other planets, revolves around the sun
7. the apparent 'stations and retrogressions' of the planets are due to the same cause.

The concept of a sun-centered universe wasn't his own idea; he was familiar with the work of Aristarchus. Copernicus's heliocentric universe still needed Ptolemy's epicycles for perfect circular orbits to match the observations, but managed to make the system work with 'only' 34 circles (as opposed to Ptolemy's 80). He published a small hand-written summary, titled *Commentariolus,* in about 1514 and circulated it among friends and a few colleagues.

He worked out the mathematical details, and by 1530, completed the manuscript of the *Book of Revolutions* - then locked it in his drawer for almost 14 years, not permitting publication until the year he died.

At about 1539, a small summary of this book was published by an associate of his, Rheticus, anonymously, at Copernicus's request.

The final manuscript was published in 1543 and was not put on the Index (a list of books banned by the Catholic Church) until 1616 when Galileo stirred up controversy with the Aristotelians and some of the Jesuit priests whom he enjoyed taunting. The book remained on the Index for only four years. After the removal of a few sentences which implied that the Copernican system was a proven theory, it was re eased, allowing people to read, discuss and teach Copernicus.

Tycho Brahe (1546-1601)

As Copernicus has been called the most colorless of the great scientists of the Renaissance, Tycho Brahe was probably the most colorful. At least, he had a golden nose - a prosthesis reputed to be made of gold and silver alloy after his own was sliced off in a duel over mathematical ability.

I was always fascinated by what motivates great scientists to devote the time and energy required for exceptional accomplishments. In Brahe's case, it was a solar eclipse in 1560 that had been predicted prior to the event. He was flabbergasted by the power of astronomy to forecast such an incredible occurrence.

Another event helped to determine his specialty within his chosen field. In 1563, he observed a close approach of Jupiter and Saturn, and to his outrage and disappointment, it was a month later than the predicted date. He decided that he could do better and, from then on, observational precision became his passion, bordering on obsession.

Luckily, he was also rich and had rich patrons, such as King Frederick II of Denmark and Rudolph II, Holy Roman Emperor, so he spent the rest of his life in luxury - twenty years of it on his own island, complete with castle and servants. He refined astronomical observations to a degree of accuracy never before dreamt of.

This turned out to be a great help to Kepler, who came to the island in 1600 as Brahe's assistant. Kepler desperately needed the best astronomical data available to complete his own theory of the Cosmos. This was a match made in heaven: Kepler, a brilliant theoretician without good data, nor access to expensive equipment, Brahe a mediocre theoretician with the best data and instruments on Earth.

When Brahe died of a burst bladder (he ate too much, drank too much, and was too polite to stand up from a baron's table to go to the washroom) Kepler, without formal leave, appropriated Brahe's accumulated observations of the past thirty years and set out to develop his theories.

Brahe's own theory was ingenious, although totally false, conciliation between the cosmologies of Ptolemy and Copernicus: he let all the planets, except Earth, orbit around the sun and then suggested that the whole menagerie rotated around Earth. In the fateful years of Galileo's trouble with the church, many well-intentioned people recommended adopting Brahe's system as a peaceful compromise.

Neither Kepler nor Galileo would accept it and the truth eventually prevailed.

Johannes Kepler (1571-1630)

Kepler was one of the most endearing and honorable of all scientists. He is the true "Sleepwalker" in the Arthur Koestler sense - bumbling through his life, pursuing a romantic and silly dream that could never come true but, as a side effect, accomplishing great progress in astronomy: sound science created for unsound reasons.

His major contributions to science are his three planetary laws. These went almost unnoticed, certainly not considered of much value by Kepler himself. Arguably, Newton could never have accomplished his great synthesis without Kepler's laws. But Newton had to dig deep for them – in Koestler's words:

"Not the least achievement of Newton was to spot the Three Laws in Kepler's writings, hidden away as they were like forget-me-nots in a tropical flowerbed. … the three Laws are the pillars on which the edifice of modern cosmology rests; but to Kepler, they meant no more than bricks among other bricks for the construction of his baroque temple, designed by a moonstruck architect. He never realized their real importance."

Kepler's 'baroque temple' was his idea that God had designed the cosmos around the five perfect solids. As Euclid had proved, there are only five perfect (all faces identical) three-dimensional geometric objects that can both fit inside a sphere with all vertices touching and encompass a sphere with all sides touching. These are:

- Tetrahedron   - four equilateral triangles
- Cube   - four squares
- Octahedron   - eight equilateral triangles
- Dodecahedron   - twelve pentagons
- Icosahedron   - twenty equilateral triangles

One day, teaching his geometry class, Kepler stopped dead in the middle of a sentence. He realized that there are exactly five perfect solids, no more, no less, and exactly (as far as he knew) six planets: Mercury, Venus, Earth, Mars, Jupiter, and Saturn. There had to be a reason for this and, in his mind, the only conceivable reason was the Grand Design of God, the skeletal structure of the Planetary System, where the spheres containing the planetary orbits (assumed at the time to be perfect circles) would be fixed in place by the perfect solids nested one inside the other.

A quick calculation showed him that the scheme had promise: into the orbit of Saturn, he fitted a cube, inside that a sphere for Jupiter; inside that a tetrahedron, inside that a sphere for Mars; inside that a dodecahedron, inside that another sphere for Earth, inside that an icosahedron, inside that a sphere for Venus; inside that an octahedron, inside that a sphere for Mercury. The known values for the planetary orbits seemed to fit (more or less) the arrangement.

Kepler was 26 years old when this notion struck him, and he spent the rest of his life pursuing the dream. He could never make it work, even by using the most precise observations of the times, Tycho Brahe's, and the most rigorous mathematical tools, many of them invented by himself. It could not work because it was utter nonsense.

What proves Kepler so honorable is his stubborn refusal to cheat a little, as was often done by his colleagues. He discarded results on which he had labored for years, just because there was a small discrepancy between his theoretical predictions and the observed facts.

During this long and fruitless search for God's Grand Design, he discovered his three planetary laws:

> **Law I.**
> The orbits of the planets are ellipses with the sun in one of the focal points
> **Law II.**
> The planet to sun radius sweeps out equal areas in equal time periods.
> **Law III.**
> The squares of the periods of any two planets are proportional to the cubes of their mean distances from the Sun.

The First Law was a major improvement over the Aristotelian obsession with perfect circles that had made all those crazy epicycles necessary. Kepler did not like the ellipse; he considered it ugly - but less so than epicycles. As he said later: "I have cleared the Augean stables of astronomy of cycles and spirals and left behind me only a single cartful of dung".

The Second Law describes one planet's - any planet's - motion around the sun in precise mathematical detail. It basically says that the planet speeds up as it approaches the sun and slows down as it moves farther away. This suggested to Kepler (and Newton) that the sun had something to do with the way the planets moved - the sun was somehow controlling the speed of their motion. This was a necessary hint before a gravitational force could be envisioned.

The Third Law compares the five planets with one another and finds that the farther they are from the Sun, the slower they move - in a precise mathematical proportion - which also suggests that the sun has a major influence on the planets.

Kepler didn't quite understand inertia, even though he was talking about the 'laziness' of the planets that makes them lag behind spoke-like forces that emanate from the sun and, like invisible brooms, sweep the planets around. However, he almost came to state universal gravity, when he recognized that:

"Gravity is the mutual bodily tendency between cognate bodies towards unity or contact.....so that the Earth draws a stone much more than the stone draws the Earth......If the Earth and the moon were not kept in their respective orbits by a spiritual or some other equivalent force, the Earth would ascend towards the moon one fifty-fourth part of the distance, and the moon would descend the remaining fifty-three parts of the interval, and thus they would unite. ...." (*Astronomia Nova,* 1609)

This is as close as you can get without actually stating the assumption of "universal gravitation" as Newton did. I am sure Newton was helped by this paragraph; he must have come across it while reading Kepler's work.

Galileo Galilei (1564-1642)

Galileo's major contribution to science was his theory of motions on Earth. He stated the concept of inertia for bodies moving on Earth correctly and announced the superposition principle. This describes the trajectory of projectiles as a parabola resulting from the forward, straight-line, uniform-speed inertial motion and the vertically accelerating (by the square of time) motion of free-fall. These concepts were of great help to Newton when he tried to explain the elliptical orbits of the planets.

In Galileo's words, from his *Dialogues Concerning Two New Sciences* (Fourth Day):

"Imagine any particle projected along a horizontal plane without friction; then we know, from what has been more fully explained in the preceding pages, that this particle will move along this same plane with a motion which is uniform and perpetual, provided the plane has no limits. But if the plane is limited and elevated, then the moving particle, which we imagine to be a heavy one, will on passing over the edge of the plane acquire, in addition to its previous uniform and perpetual motion, a downward propensity due to its own weight; so that the resulting motion which I call projection, is compounded of one which is uniform and horizontal and of another is vertical and naturally accelerated. We now proceed to demonstrate some of its properties, the first of which is as follows:

Theorem I, Proposition I: A projectile which is carried by a uniform horizontal motion compounded with a naturally accelerated vertical motion describes a path which is a semi parabola."

Galileo has firmly established the scientific method by replacing the often baseless Aristotelean speculations with rigorous experiments, observation and measurements.

He states:

"Philosophy [nature] is written in that great book which ever lies before our eyes. I mean the universe, but we cannot understand it if we do not first learn the language and grasp the symbols in which it is written. The book is written in the mathematical language, and the symbols are triangles, circles, and other geometrical figures without whose help it is humanly impossible to comprehend a single word of it, and without which one wanders in vain through a dark labyrinth." *The Sidereal Messenger* (1610).

and in his *Discourses and Mathematical Demonstrations Concerning Two New Sciences* (1638) he says:

"The cause of the acceleration of the motion of falling bodies is not a necessary part of the investigation."
        .....
"Nature nothing careth whether her abstruse reasons and methods of operating be or be not exposed to the capacity of men. . .. When we have the decrees of nature authority goes for nothing."

Galileo's contribution to astronomy was, surprisingly, modest. Contrary to popular belief, he did not invent the telescope. The earliest record of an existing telescope is from a patent application in

Holland on 2 October 1608. When Galileo heard about the invention, he acquired his own and started using it for astronomical observations. After he discovered Jupiter's moons, Venus's phases and finally stood up for the Copernican system, he accomplished little and committed serious blunders.

-      He dismissed Kepler's Laws and elliptical orbits
-      He ridiculed Kepler's correct explanation of the tides
-      and provided a totally false one himself.
-      He stated that the natural motion of the planets was perfect circles.
-      He accepted Copernicus's Epicycles - all 34 of them.

Two of these assertions deserve comments:

1. Even though Kepler accurately stated that the cause of the tides is the attraction of both the sun and the moon, Galileo thought it was a ridiculous "astrological superstition" and suggested that the tides were the result of the difference between the speed of the Earth's surface on the side facing the sun and the side facing away from the sun.

2. It really surprised me when I first learned about Galileo's suggestion that the natural motion of the planets is perfect circles. Up to that point, I had assumed that Galileo discovered the principle of inertia, as is commonly stated in most physics textbooks. And, while it is true that Galileo stated a principle of inertia in his later work *On Two Sciences* as valid on Earth, apparently he did not think this principle applied to heavenly bodies, as he very clearly described it in his *Dialogue,* published just a few years earlier. The *Dialogue of two World Systems* is written in the form of a debate between two fictional characters: Salviati, representing Galileo's convictions, and Simplicio, representing the opposing view. Galileo reasons incorrectly:

"**SALV.** This principle being established then, it may be immediately concluded that if all integral bodies in the world are by nature movable, it is impossible that their motions should be might, or anything else but circular; and the reason is very plain and obvious. For whatever moves straight changes place and, continuing to move, goes ever farther from its starting point and from every place through which it successively passes. If that were the motion that naturally suited it, then at the beginning it was not in its proper place. So then the parts of the world were not disposed in perfect order. But we are assuming them to [be] perfectly in order; and in that case, it is impossible that it should be their nature to change place, and consequently to move in a straight line."

Please observe how Galileo forgets about his own stated principle, according to which "Nature nothing careth whether her abstruse reasons and methods of operating be or be not exposed to the capacity of men". He is theorizing in the true Aristotelian style, without any experimental observation whatsoever. He is making assumptions that are undefined and unwarranted.

He goes on in a similar vein: "Besides, straight motion being by nature infinite (because a straight line is infinite and indeterminate), it is impossible that anything should have by nature the principle of moving in a straight line; or, in other words, toward a place where it is impossible to arrive, there being no finite end. For nature, as Aristotle well says himself, never undertakes to do that which cannot be done, nor endeavours to move whither it is impossible to arrive. "

Now he appeals to the same Aristotle's authority that he ridiculed on other occasions. No, this definitely is not the Galileo we learned about in high school. It appears as if his genius worked unhindered as long as he studied phenomena on Earth, but when it came to the heavens, he reverted to superstitious speculation.

Rene Descartes (1596 – 1650)

Rene Descartes is best known for his famous sentence: *"Cogito ergo sum"* ("I think therefore I am"). This is from his *Discourse on the Method* [of Rightly Conducting the Reason and seeking Truth in the Sciences] published in 1637, a book attempting to build up a perfectly rational philosophy based on an attitude of "doubting everything". He is also the inventor of the "Cartesian co-ordinate system", named after his Latinized name: Renatus Cartesius.

His greatest contribution to science was in mathematics. He invented analytical geometry, which laid the foundation for Newton's and Leibniz's work.

There is no telling what gives a 'nerd' ideas. Lying in his bed one day, Descartes was watching a fly. The room was sparsely furnished; all the edges where the walls, ceiling, and floor met were quite visible, and suddenly Descartes realized that he could describe the position of the fly, anywhere in the room, by a set of three numbers: the fly's distance from the floor and from each of two connecting walls.

With this simple thought, he connected algebra and geometry, which had up till then been two utterly different branches of mathematics. Building on this simple yet powerful idea, he worked out a method by which difficult problems in geometry could be solved with algebra and vice versa. So, when Newton began to think about the acceleration of moving bodies, he had the foundation for inventing calculus: a necessary tool to develop his theory.

Descartes's astronomical studies were modest. In 1633, he had nearly finished a book on the ideas of Copernicus, of which he approved. Upon hearing of Galileo's trouble with the Church, he suppressed it. Instead, he developed another compromise, wherein all space was filled with matter arranged in rotating vortices. The idea of the vortices was necessary to account for the circular movements of the planets since Descartes vehemently rejected the concept of 'action at a distance'. This theory appealed to many scholars but had no scientific merit and was soon abandoned in favor of Kepler and

Newton. Newton devoted many pages in his *Principia* to discredit Descartes' vortex idea and went ahead with his theory of universal gravity.

On the positive side, Descartes was the first to state clearly that the natural uninfluenced motion of heavenly bodies is a straight line, at a uniform speed. Thus he gave the clue to Newton that Galileo's Earth-bound inertia applied to planets as well.

Chapter III – Newton's Laws

Newton's Time

In the late Renaissance, from the era of Roger Bacon and Thomas Aquinas to Newton and beyond, science was 'in the air', so to speak. It was fashionable to understand and discuss science. Educated aristocracy and the emerging middle class welcomed news of discoveries and were proud of their country's geniuses. One particular example was the 'war' between Germany and England, regarding the question of priority in inventing mathematical calculus. Both Leibniz and Newton discovered it independently, but the admirers of each accused the other of plagiarism. Contemplating this nationalistic rivalry, it is quite obvious how much scientific accomplishments were valued at the time.

Revolutionary discoveries were already in circulation in many forms before Newton actually pulled them all together in his great work: *the Principia* [The Mathematical Principles of Natural Philosophy].

A summary of the main parts of the puzzle Newton faced:

	Inertia	Weight	Force	Center	Orbits
Aristotle	Moves only when pushed	Desire of bodies	Direct contact only, spirits moving Planets	Earth	Perfect Circles, uniform speed, Epicycles
Ptolemy	Moves only when pushed	Desire of bodies	Direct contact only	Earth	Perfect Circles, almost uniform

					speed, Epicycles
Copernicus	Uniform circular	Tendency of bodies for sphere	No force needed for planets	Sun	Perfect Circles, uniform speed, Epicycles
Brahe	Uniform circular			Earth & Sun	Perfect Circles, Epicycles
Kepler	'Laziness' of Planets	Mutual attraction	Spokes of force from Sun	Sun	Ellipse with Sun in focus
Galileo	Uniform circular	Absolute quality of bodies	No force needed for planets	Sun	Perfect Circles, uniform speed, Epicycles
Descartes	Uniform straight line		Vertices in ether	Sun	Perfect circles

We have now seen the ideas of intelligent and accomplished scholars of the time. No wonder it required a genius of Newton's caliber to cut through this maddening confusion.

Isaac Newton (1642 - 1727)

Newton had a modest start in his academic career. According to James Gleick (*Isaac Newton*):

"His mother supplied him with a chamber pot; a notebook of 140 blank pages... a quart bottle and ink to fill it, candles for many long nights, and a lock for his desk".

Try sending your child to university so equipped today.

Unlike Kepler with his enthusiasm, openness, desire to communicate his innermost thought processes, Newton was a vain, insecure, aggressive, and very vindictive genius.

At this point, I have a problem with my narrative. How do I make an un-likable man interesting, fascinating, awe-inspiring? Human emotions, being what they are, often prevent us from an objective evaluation of people we don't like.

There are geniuses who are easy to 'sell' to the public. For example, everyone's favorite: Johannes Kepler: honest, open, generous, romantic, hardworking, self-mocking, enthusiastic, and brilliant – how can one not like him?

Or Tycho Brahe, the 'enfant terrible' of Astronomy, the spoilt aristocrat, the high-living lord on his own island with his own castle, with his own subjects, with his unmatched precision in observation?

Or Richard Feynman, the bongo-drum-playing prankster with the charm of a curious child, the soul of a poet, and a mind like a laser beam?

We like colorful geniuses, people just like us, only much smarter. Carl Sagan devoted a segment of his epic *Cosmos* documentary series to Kepler and Brahe, while he barely mentioned Newton. Newton wasn't like them. He was dry, humorless; he lived a life so boring that no interesting scandal ever corrupted his name. He never married, never showed an interest in women; his only hobbies besides science were alchemy and theological speculation.

Newton did not tell all what he had in his mind or how it got there: he kept delaying publication of his books, often to his own disadvantage, until he had them worked out to perfection. The least hint of criticism or even doubt sent him into rages, even bordering on nervous breakdown. He carried on vicious, vindictive feuds for decades, over slights he mostly imagined, and threatened to abandon science altogether, due to lack of proper admiration and respect.

Yet, reading the *Principia* for the first time is an awe-inspiring experience. Published in 1687, three hundred and twenty-seven years ago, it is still modern and fresh today. Newton spent a little over two years writing this precise, complete, fundamental, and powerfully reasoned treatise. He convinced his contemporaries that all problems could be solved by rigorous and logical thinking.

This optimism is probably Newton's greatest legacy: man's confidence in himself and his ability to understand, and cope with, Nature.

One can feel fascination and awe when confronted with a mind like Newton's. His genius is part of what we are; we would be the poorer without knowing of him. I often imagine him looking intently at his manuscript and wonder whether the paper ever burst into flames from the focussed, concentrated beam of his intellect.

The power of Newton's theory is that it explains everything concerning motion and forces. Not only falling bodies, moving planets, accelerating automobiles - it also applies to seemingly unrelated phenomena such as sound, heat, gases, liquids, turbulence, vibration, stress, deformation, and elasticity. It is used today in calculating design parameters for industrial machinery, cars, ships, submarines, airplanes and space vehicles; in skyscrapers, bridges, canals, cranes, and other construction equipment. The consequences of a few definitions and three basic laws are staggering.

Until electromagnetism, relativity, nuclear, and quantum physics were discovered, Newton's theory explained all of the known universe.

And now, it is time to dive into Newton's theories.

I have to warn you that the colorful descriptions are over. We have to change our tone from that of entertaining anecdotes to serious, penetrating intelligence.

This is a challenge I like to pose for myself. If one human being was able to conceive these ideas, to invent most of the concepts, to see and prove the hidden underlying connections that had stumped all of his predecessors and contemporaries, then am I smart enough to at least understand when he explains it so clearly?

Because, my friends, if we don't learn from those who solved nature's mysteries, we will not master the tools we so desperately need to save our civilization from self-destruction. We can't have it both ways! We can't enjoy continuing affluence while discarding, often despising, the thought process needed to create and maintain it. There is no substitute for careful observation, precise, well-defined stating of objectives, intelligent analysis, theorizing, and predictions. If we want predictable results that work in areas of non-science (like our collective survival) then we have to learn to think as great scientists taught us to think.

Newton's Laws

We know very little about how Newton arrived at his synthesis – we only know that he had worked on laws of motion and gravitational attraction on and off through most of his adult life. I have sketched out the story of the partial discoveries on which Newton was building, but we will never know how much of which predecessors influenced him, inspired him, helped him along with his project.

Robert Hooke, Newton's archrival, claimed that "universal gravity" was *his* idea. We have copies of letters written by Hooke, telling Newton about the concept and asking his opinion, but we will never know whether the concept had already been clear in Newton's own mind when he received those letters. As he stated in a letter to Hooke: "If I have seen farther, it is by standing on the shoulders of giants."
(Contrary to general belief, this was not a statement of humility, but a clever way of rubbing Hooke's nose in Newton's presumed superiority)

Whoever had priority, the important fact is that the synthesis gave such a boost to the cause of science that almost all opposition from the church and the Aristotelian academia disappeared. Scientists were free to pursue the study of nature along the path of observation, measurement, hypothesis, and mathematical deduction, as laid out by Kepler and Galileo. Newton provided them with a consistent worldview that could explain:

- all the observed motions of planets and the sun
- all the observed movements of the moon
- the phenomenon of the tides

- the observed trajectory of comets
- all observed movement of projectiles on Earth

The essence of Newtonian mechanics is the realization that:
- The same laws work both on Earth and in the heavens.

- Inertia is an innate property of material bodies and operates in a straight line and with uniform speed.

- Planetary orbits are the result of the superposition of the force of inertia and the force of gravitational attraction, acting on the body together.

- The gravitational forces are universal, acting not only between heavenly bodies but between any two material particles possessing mass.

- The motion through space and time of any material body can be calculated if we know the mass of the body and the forces acting upon it.

I will quote quite a bit from Newton's original manuscript because it is almost impossible to appreciate his accomplishments without hearing his own words. Gone are the confusing assumptions in Copernicus's book, the mystic references in Kepler's writings, the baseless speculations from Galileo's text on Astronomy. Newton is focussed, precise as a surgeon's scalpel; absolutely nothing is left to the readers' imagination, nothing left undefined. Newton's *Principia* is the first solidly forged chain of scientific logic, awesome in its completeness. He even attempted to define time and space, but he only touched on the topic to show that he was aware of their significance – further progress had to wait for Einstein.

Modern science is built on the foundation of precise definitions of basic terms. These definitions can be conceptual or practical (like measuring instructions) but essential in order to forge the unbroken chain that connects theories to fundamental experience. Before Newton formulated his laws, he defined his basic concepts as follows:

Definition I
"The **quantity of matter** is the measure of the same, arising from its density and bulk conjunctly".
Definition II
"The **quantity of motion** is the measure of the same, arising from the velocity and quantity of matter conjunctly".
Definition III
"The VIS INSITA, **or innate force of matter**, is a power of resisting, by which everybody, as much as in it lies, endeavours to preserve in its present state, whether it be of rest, or of moving uniformly forward in a right line".
Definition IV
"An **impressed force** is an action exerted upon a body, in order to change its state, either of rest, or of moving uniformly forward in a right line".

In **Definition I** Newton introduced the concept of **mass** as the *quantity of matter.* He referred to the common experience of finding mass proportional to the density and volume of the material body in question.

In **'Definition II'** he introduced a concept that we call today **momentum,** that is mass times velocity of a given object. Newton called it *quantity of motion*, recognizing that the two essential ingredients of motion were the mass and the speed of the moving object.

In **'Definition III'** he defined **inertia** as an innate force of matter that makes material bodies resist acceleration (change of speed and or direction) of any kind. He did not explain why this 'force' exists, merely recognized its existence.

Finally, **'Definition IV'** describes **force** as the primary cause for acceleration; wherever that force originates, it must be present if a body is changing speed and/or direction.

After these definitions, Newton was ready to state his laws as follows:

Law I
"Every body perseveres in its state of rest, or of uniform motion in a right line, unless it is compelled to change that state by forces impressed thereon"
Law II
"The alteration of motion is ever proportional to the motive force impressed; and is made in the direction of the right line in which that force is impressed'
Law III
"To every action there is always opposed an equal reaction: or the mutual actions of two bodies upon each other are always equal, and directed to contrary parts"

"Big deal", I thought when I learned this, "sounds pretty obvious to me".

The first law seemed obvious, anyway. Early experience on the skating rink taught me that the smoother the surface, the longer an object slides on it without changing speed or direction. I did not know why objects had this property of inertia but accepted it as a fact of nature.

The third law also seemed obvious. Think of a cat jumping off a chair and kicking the chair back at the same time, or two people on skates, one pulling the other towards him, but both moving at the same rate and meeting somewhere in the middle. You can't help seeing the law in action.

It was the second law I had a problem with. It states, again: "The alteration of motion is ever proportional to the motive force impressed". By "The alteration of motion" Newton meant the deviation from the straight-line and uniform-speed inertial motion, which is the natural state of objects, according to Law I. This deviation is what we call the acceleration of a material object, so Newton's second law basically says that force equals mass times acceleration. Transcribing it into mathematical form

$$F = m_i \, a$$

where '**F**' is the force acting on the body, '**a**' is the acceleration of the object (which we know by measuring distances, angles and time) 'm_i', is the **quantity of matter** (pronounced as m-sub-i) where the little subscripted 'i' means 'inertial'. We call it the body's **inertial mass**, because it depends entirely on the inertia (resistance to acceleration) of the object (how much force we have to exert on it to make it accelerate at '**a**' acceleration). This is the object's mass arising from its density and bulk conjunctly, according to Definition I. Newton assumed the quantity of matter to be a permanent attribute of material bodies, only dependant on density and volume (bulk).

This assumption proved to be incorrect as Einstein pointed out in 1905 in his "Special Theory of Relativity" because 'm_i' depends also on the speed of the body moving relative to an observer, but this effect is not easily noticeable until the speed reaches a substantial fraction of the speed of light: 300,000km/sec. At lower speeds this effect is negligible, so we will ignore it for the time being.

Here was my problem.

We have an equation with three values in it. I know one of these three: I know what acceleration is and how to measure it. I have an instinctive 'feel' for the concepts of mass and force, but I have no idea how to measure them, and without measuring instructions, no concept in physics has any practical value.

How do we measure this **inertial mass:** m_i and **force:** F?

Any time I asked my teachers what mass was, I was told that it is the object's inertia (force divided by acceleration). And when I asked what force was, I was again told, predictably, that it was mass times acceleration.

Finally, years later, I read a few books that satisfied my curiosity. One of the best: Richard Feynman's *Lectures on Physics* recognizes the validity of my question:

"If we have discovered a fundamental law, which asserts that the force is equal to the mass times the acceleration, and then define the force to be the mass times acceleration, we have found out nothing....the physical law F=ma is an incomplete law."

When I read this, my philosophical problems were solved: Now I knew what Newton meant and how to use his law to solve practical problems. Once I substitute the formula for the force, be it gravity, tension in springs, etc., I can easily determine the motion of any object subjected to the force if I know the mass of the object.

Usually, mass is measured by weighing things, but this is not entirely correct. Weighing measures the force of gravitational attraction between the Earth and the body being weighed (gravitational mass). This is conceptually different from the 'inertial mass', which is defined as an object's resistance to acceleration. To make it more clear: just think of how the weight of an object will drastically change (to about one-sixth of its value on Earth) if measured on the moon, while its inertial mass - resistance to acceleration - will remain the same.

We are not completely defeated, though, because we have ways to measure inertial mass without weighing the object, by measuring the acceleration of objects under identical forces: e.g. accelerating them via the same spring stretched to the same extent. If the force is identical on two objects with masses m_{i1} and m_{i2}, then $F=m_{i1}a_1= m_{i2}a_2$, which means that

$$m_{i1}/ m_{i2}= a_2/ a_1.$$

This means that we can measure the *ratio* of inertial masses between any two objects by measuring their accelerations under identical forces and, if we arbitrarily select one particular object as a unit for inertial mass, then we can measure the inertial mass of any object as described above (the prototype of 1 standard kilogram is a cylinder made of platinum-iridium alloy, kept at the International Bureau of Weights and Measures in Paris).

Now that we have a unit of inertial mass we can define the unit of force as well: 1 N (one **newton**) is the amount of force that causes one unit of inertial mass (1 kg) to accelerate at 1 meter per second per second. Thus, we have a way to measure both inertial mass and force in their respective units.

Of course, measuring masses by accelerating them with springs is cumbersome and fortunately, we don't have to. Instead, as we will soon see, we can take advantage of what seems at first a lucky coincidence: the inertial and gravitational masses of objects happen to be the same, so we can measure both masses by weighing the objects, after all.

Universal Gravity

Newton's three Laws would be of little value had he not given us the concept and the formula for the force of universal gravitation.

Newton arrived at this formula by analyzing Kepler's second and third laws, especially the third law, which explicitly uses the distance of a planet from the sun as a measure of the sun's 'power' over the planet. However, Newton, with powerful insight, suggested that this force applies to **any** two material objects.

In his *Principia*, he makes it very clear:

> **Proposition VII. Theorem VII.**
> That there is a power of gravity tending to all bodies, proportional to the several quantities of matter which they contain.

The proposition says that a material body has an attribute (**gravitational mass**) that responds (by being attracted) to the same attribute in another material body and that this attribute depends on the quantity of matter the affected bodies contain (bulk and density) – that is, their masses. Gravitational mass is not necessarily the same as **inertial mass** (as defined above). Only precise experimental observation can determine how they are related.

The idea of universal gravitation seemed utterly ridiculous to most of Newton's contemporaries. To suggest that heavenly bodies exert some kind of force upon each other at a distance was odd enough, but a case could be made for that assumption when the orbits of the planets were analyzed. On the other hand, the idea that ordinary objects here on Earth attract each other with a mysterious force seemed unfounded. Nobody ever observed such a force between, say, my ink-well and the candle stick.

Still, Newton was convinced that such a force did indeed exist, and he countered his critics with corollary No. 1

Cor 1.

…all bodies with us must mutually gravitate one towards another, whereas no such gravitation anywhere appears, I answer, that since the gravitation towards these bodies is to the gravitation towards the whole Earth as these bodies are to the whole Earth, the gravitation towards them must be far less than to fall under the observation of our senses.

That is: since the masses of ordinary objects on Earth are so tiny compared to the mass of Earth, the forces between them must be extremely small as well, compared to the force exerted on each of them by the Earth, and that is why we have never seen any evidence of these forces. (We will soon see a description of the first experiment that demonstrated them.)

Newton went beyond stating the existence of universal gravity: he gave us the exact formula for how this force depends on the masses of the objects and their distances from each other. In **_Proposition VII. Theorem VII._** he had already stated that the force is proportional with the masses of the objects, and in corollary No. 2 he tells us how the force also depends on the distance of the objects:

Cor.2

The force of gravity towards the several equal particles of any body is reciprocally as the square of the distance of places from the particle

---

The formula is very simple: it says that between any two material objects, there is a mutually attracting force, directly proportional to the gravitational masses of the two objects and inversely proportional to the square of their distance.

$$F = G \, m_{g1} \, m_{g2}/d^2$$

where 'G' was an as yet undetermined proportional constant factor, to be determined by measuring the force between two objects of unit gravitational masses at unit distance from each other (if $m_{g1} = m_{g2} = 1$ and $d=1$ then $F=G$).

Inertial mass and gravitational mass

The unit of gravitational mass has not yet been chosen, but it will be, and then it will determine any object's gravitational mass from Newton's formula: if the gravitational attraction is twice as strong on an object as on the unit, then it is assumed to have a gravitational mass of 2 units.

Let's consider a material object of inertial mass m_i and gravitational mass m_g falling freely in a gravitational field $\mathbf{F_g}$ (F_g is the gravitational force exerted on an object of one unit of gravitational mass).

Since F_g is the gravitational force exerted on one unit of gravitational mass, the force exerted on an object of m_g units of gravitational mass must be

$$F = m_g \, F_g$$

According to Newton's Second Law $F = m_i g$ ('g' is the free-fall acceleration in the gravitational field. Therefore:

$$g = F / m_i$$

Substituting F from the first formula above:

$$g = (m_g / m_i) \, F_g$$

That is the free-fall acceleration in a given gravitational field depends only on the m_g / m_i ratio of material objects.

According to countless experiments (started by Galileo), the free-fall acceleration for any object in Earth's gravitational field (close to Earth's surface) is the same: $g = 9.81$ [m/s^2] which means that the (m_g / m_i) ratio is the same for all material objects on Earth. It is a reasonable assumption that this holds true in any gravitational field, not just Earth's.

What remains to show now is that this ratio equals 1 if we choose the same unit for both inertial and gravitational mass.

This experiment was first performed by the Hungarian Lorand Eotvos in 1909 and later by other physicists. The results show that the 'gravitational mass' and 'inertial mass' of a material object is the same value within an error limit of 1 part in 1,000,000,000. Therefore, we can say that an object's inertial mass and gravitational mass are the same, so from now on, we will simply talk about an object's mass.

If we look back at our definition of an object's inertial and gravitational mass, we see that the identity of the two means that:

the attribute of a body that makes it resist acceleration is the same attribute as the one that responds to gravitational attraction.

This is a deeply philosophical, and by no means obvious conclusion, and will be used as one of the pillars on which Einstein built his General Theory of Relativity.

One more consequence: since the **weight** of an object on Earth is the force of gravitational attraction of Earth on the object, now we can measure the inertial mass of an object by weighing its gravitational mass on a scale suitably designed for Earth. On the moon, we would need different scales, because the force of attraction on the moon is about one-sixth of its value on Earth. This will cause a gravitational acceleration proportionally smaller as well because the 'gravitational mass' of an object is the same everywhere.

Calculating the masses of Earth, Moon, Planets.

As stated above, the formula for universal gravitation, as given by Newton is: $F = G\, m_{g1}m_{g2}/d^2$ where 'G' is an as-yet undefined proportional constant factor, to be determined by measuring the force between two objects of unit gravitational masses at unit distance from each other (if $m_{g1} = m_{g2} = 1$ and $d = 1$ then $F = G$).

When the unit of gravitational mass was selected (1 kg, same as the unit of inertial mass), it became possible to measure the attraction between two objects on Earth and thus determine the value of the gravitational constant 'G'.

The experiment was first performed by **Henry Cavendish in 1798** in an ingenious way. He suspended a rod, with identical small lead balls on its ends, by a wire attached at its center. He suspended two large balls of identical mass close to the two small balls, one to either side of the rod, so that the centers of small and of the large balls were equidistant from the wire. He then measured how much the wire twisted due to the gravitational attraction between the small and large balls. The amount of the twist gave him the force of attraction.

Since he knew the masses of the balls and the distance between them, using the gravitational formula (only 'G' was unknown) he calculated the value of 'G' as

$$G = 6.670 \times 10^{-11}\ [\text{newton} \times \text{meter}^2/\text{kg}^2]$$

Once G was known, we could easily measure the mass of Earth by measuring the gravitational acceleration of an object on the surface of Earth. Newton's gravitational formula, applied to Earth and an object of mass 'm', states that if the mass of Earth will be represented by 'M', the radius of Earth by 'd' and the free-fall acceleration by 'g' then the force acting on the object is:

$$F = GmM/d^2 = mg$$

from which

$$GM/d^2 = g\ ,$$

from which

$$M = g \, d^2/G$$

We know from countless measurements that the value of freefall (in vacuum) acceleration of any objects close to the surface of Earth is

$$g = 9.81 \, [m/s^2]$$

We also know the radius of Earth (distance from the surface to center) as

$$d = 6,370,000 \, m$$

and we just obtained the value for 'G' (see above), so the calculation of the Earth's mass gives us the value of

$$M = 5.97 \times 10^{24} \, kg$$

Since this value would be impossible to calculate without the result of Cavendish's experiment, physicists, being only human, dubbed it the "Weighing the Earth" experiment.

Once we know the mass of the Earth, then we can determine the mass of the moon and other planets from the same gravitational formula, the fully documented orbital parameters, and some simple arithmetic.

What is gravity and how does it operate?

Newton had no explanation for gravity: a force acting instantly, through millions and millions of miles of vacuum, whipping about objects (Earth, moon, planets) of incredible size and mass, without any material contact whatsoever, as if they were little pebbles at the end of a child's string. A steel cable the diameter of the Earth itself, would not be strong enough to hold Earth in its orbit. It would snap under the strain of centrifugal force.

Many contemporary scientists rejected the idea as a regression to Aristotle's spiritual movers. Newton himself refused even to speculate about the nature of gravity, as he acknowledged in his *Principia: (Book III General Scholium)*

"Hitherto we have explained the phaenomena of the heavens and of our sea by the power of gravity, but have not yet assigned the cause of this power. This is certain, that it must proceed from a cause that penetrates to the very centers of the sun and planets, without suffering the least diminution of its force; that operates not according to the quantity of the surfaces of the particles upon which it acts (as mechanical causes use to do), but according to the quantity of the solid matter which they contain, and propagates its virtue on all sides to immense distances, decreasing always in the duplicate proportion of the distances ....But hitherto I have not been able to discover the cause of those properties of gravity from phaenomena, and I frame no hypotheses"

Privately, he expressed his doubts in much stronger words:

"It is inconceivable that inanimate, brute matter should, without the mediation of something else, which is not material, operate upon, and affect other matter without mutual contact... And this is one reason why I desired you would not ascribe innate gravity to me. That gravity should be innate, inherent, and essential to matter, so that one body may act upon another, at a distance through a vacuum, without the mediation of anything else, by and through which their action and force may be conveyed from one to another, is to me so great an absurdity, that I believe no man who has in philosophical matters a competent faculty of thinking, can ever fall into it. Gravity must be caused by an agent, acting constantly according to certain laws; but whether this agent is material or immaterial, I have left it to the consideration of my readers. " Quoted from letter to Richard Bentley.

And this is where the theory of gravitation remained until 1915 when Albert Einstein published his "General Theory of Relativity", in which he gave us a (non-Euclidean) geometrical interpretation of gravity as the curvature of space-time caused by the presence of mass. The concept of curvature of space-time is so esoteric, non-intuitive, that very few laymen understand it accurately. Simon Singh tells us an amusing story about Arthur Eddington in his excellent book: *Big Bang*

"Eddington became so closely associated with the theory that the physicist Ludwig Silberstein, who also considered himself an authority on general relativity, once said to Eddington: 'You must be one of three persons in the world who understands general relativity'. Eddington stared back in silence until Silberstein told him not to be so modest. 'On the contrary' replied Eddington, 'I am trying to think who the third person is.'"

Before delving into the delightful and mind-bending subject of relativity, we have to do a lot of work: first, we must fully understand electrodynamics, 'speed-of-light experiments, and the "Special Theory of Relativity".

Space and time

All of our discussions so far were concerned with movement and its attributes: distance, speed, acceleration, orbit, etc. Unstated, we had two common-sense assumptions in discussing all of these concepts:

1.   our implied ability to measure space and time.
2.   all measurements made relative to some observer

Point 1. is obvious: We need clocks and measuring tapes. Point 2. requires a little more thought:

Why is it important to emphasize "relative to the observer"? It seems self-evident. However, if two observers are moving relative to each other - one on Earth and the other on the moon, for example -

and both are studying the same comet that moves relative to both, then their description of events will not be identical.

The question is: which will be correct? The answer: both will be, from their own perspective.

Another question: will they describe observed natural laws the same way? The answer: it depends.

Here is one very simple example. If we make our observations in a rotating room, we will think that Newton's First Law is not valid. Objects left alone do not stay at rest or proceed in a straight line at uniform speed, but fly to the walls, accelerating all the way.

One may object that the example is unrealistic: no person of sound mind would want to perform experiments in a rotating room. But we *are* doing exactly that! The Earth, on the surface of which we perform experiments, *is* a rotating spherical room and we are sitting on the *outside* of it with our laboratories and instruments. Many observed phenomena can only be explained, without contradicting Newton's theory, by taking this rotation into account.

In 1835, Gaspard de Coriolis proved that on the spinning surface of Earth, objects moving north from the equator are pushed eastward by some force, and those moving south from the equator are pushed westward. This force is due to the rotation of Earth and is now called "Coriolis force". It must be taken into account in artillery fire and satellite launching, for example. It also explains the whirling motion of hurricanes and tornadoes.

The existence of this force can be deduced from Newton's equations if they are applied to a rotating coordinate system. If we were not aware of the Earth's rotation, then we might think that Newton's equations are false because they contradict some of our experimental results!

Incidentally, the first experiment proving that the Earth *is* rotating around its axis was performed by Jean Bernard Leon Foucault in 1851. He deduced from Newton's laws that if a large pendulum were set into motion on Earth, it would maintain its plane of oscillation while the Earth rotated under it. To prove this, he suspended a two-foot diameter iron ball from a steel wire more than 200 feet long from the dome of the Paris Observatory.

As observed by an excited crowd, the pendulum's plane of oscillation slowly twisted in the direction and at the rate predicted by Newton's equation. What they actually observed was the Earth rotating under the pendulum, carrying the people and the church along, while the pendulum maintained its plane of oscillation in space.

So, as we have seen, it is important to be aware of the movement of the observer when trying to deduce the laws of nature. It is important to think about this 'relativity' because we are at the beginning of the 'space-age' when we start to travel at great speeds, covering larger distances, with fewer and fewer reference points to guide us.

Think about it! In our ordinary, everyday lives, we see motions all over the place. Cars whizzing by on the road, clouds on the sky, leaves swept by the wind, pedestrians on the sidewalk, all move in some direction at some speed. It is our natural environment and we have no problem relating to it. We have reference points against which to measure movement: the road surface, the sidewalk, the horizon; we see distances between objects increasing and decreasing, so we assume that they move relative to each other.

However, when we sit on a train, looking out the window, we may see another train on the next track moving slowly relative to us and not immediately know whether it is our train or the other that's moving on the tracks. We observe the relative movement, but not the absolute one - if there is such a thing.

The inevitable question that popped into some curious minds was: if I am floating in empty space, so far away from stars and galaxies that absolutely nothing is visible, how can I know whether I am stationary, moving ahead, or rotating?

Not until Einstein in 1905 did scientists seriously - and successfully - consider this question. When Einstein did, our understanding of reality made such a huge leap, that our 'common sense' concepts of time and space were destroyed overnight. The mysteries of Einstein's relativity principle are the first indication we had that reality is much deeper and weirder than we could have imagined before. It was a serious blow to our complacent confidence. But relativity is the topic of the next part of this book after Electrodynamics is clearly understood.

Newton was aware of this question of 'relativity' and tried to determine where and how his laws were an accurate description of nature. He started by 'dodging' the definition of 'space' and 'time'.

As he wrote in his *Principia*:

"I do not define time, space, place and motion, as being well known to all…… it will be convenient to distinguish them into absolute and relative, true and apparent, mathematical and common."

He then proceeds to define 'absolute time', 'absolute space', place, and motion:

I. Absolute, true, and mathematical time, of itself, and from its own nature flows equably without regard to anything external, and by another name is called duration: relative, apparent, and common time, is some sensible and external (whether accurate or unequable) measure of duration by the means of motion, which is commonly used instead of true time; such as an hour, a day, a month, a year.

II. Absolute space, in its own nature, without regard to anything external remains always similar and immovable.

III. Place is a part of space which a body takes up, and is according to the space, either absolute or relative.

IV. Absolute motion is the translation of a body from one absolute place into another; and relative motion, the translation from one relative place into another.

After these lofty definitions, he admits that we have to use measurements, rather than absolute knowledge, to describe our observations:

"But because the parts of space cannot be seen, or distinguished from one another by our senses, therefore in their stead we use sensible measures of them: For from the positions and distances of things from any body considered as immovable, we define all places; said then with respect to such places, we estimate all motions, considering bodies as transferred from some of those places into others. And so, instead of absolute places and motions, we use relative ones;"

Then he grudgingly admits that for all practical purposes we will have to use the measurements available:

….."for it may be that there is no body really at rest, to which the places and motions of others may be referred."

Newton holds out a glimmer of hope for us mortals: there may be a way to decide whether a body is at rest or moving at a uniform speed in a straight line:

"It is indeed a matter of great difficulty to discover, and effectually to distinguish the true motions of particular bodies from the apparent; because the parts of that immovable space, in which those motions are performed, do by no means come under the observation of our senses. Yet the thing is not altogether desperate; for we have some arguments to guide us, partly from the apparent motions, which are the differences of the true motions; partly from the forces, which are the causes and effects of the true motions."

And here comes Newton s thought experiment:

"For instance, if two globes, kept at a given distance one from the other by means of a cord that connects them, were revolved about their common centre of gravity, we might, from the tension of the cord, discover the endeavour of the globes to recede from the axis of their motion, and from thence we might compute the quantity of their circular motions. …..And thus we might find both the quantity and the determination of this circular motion, even in an immense vacuum, where there was nothing external or sensible with which the globes could be compared."

That is, we can determine whether an object is truly accelerating relative to absolute space, even in a totally empty space where there is nothing to which we can compare the movement of the body, by observing the presence or absence of forces acting on the body. If it is truly accelerating relative to absolute space, then we will see the effect of this acceleration: for example, a rope will be pulled taut between two balls; water will acquire a concave surface in a spinning bucket.

So. Newton declared that his Laws are valid in any system of observation where the system is at rest, or moving at a uniform speed, in a straight line, relative to this 'absolute space'. The consequence of this reasoning is the assumption that if the rope and the two balls are not accelerating relative to absolute space, but somehow the room in which the rope is suspended is rotated around it, the rope will remain slack, even though for the observer inside the room, who is unaware of the room's rotation, it seems obvious that the rope with the two balls are rotating around an axis.

I wanted to discuss this aspect of Newton's laws before we carried on with more practical matters because these considerations were essential for Einstein in formulating his "General Theory of Relativity". And, of course, without the theory of relativity, we would forever wonder why our spaceships had difficulty accelerating as we approached the speed of light. We are not there yet, but might someday, and then we won't want to be stuck in the middle of interstellar space because we underestimated the amount of fuel required from Earth to Tau Ceti.

The power of Newton's laws

Soon after the *Principia* was published, its power was demonstrated all over the world as scientists, applying its principles, was not only able to explain many previously unexplained phenomena, but also to solve long-standing problems.

As a first result, all of Kepler's laws can be deduced mathematically from Newton's second law and the formula for universal gravitation. The phenomenor of the tides can also be precisely explained by Newton's laws, including the mysterious fact that there are two tides simultaneously on the opposite sides of Earth, rotating around the planet every 24 hours, which produces a tide at any one spot every 12 hours. This fact had totally stumped Galileo when he tried to explain the tide from incorrect assumptions.

One of the most dramatic confirmations of Newtonian Physics was the discovery of Neptune. Analyzing the irregularities in the motion of Jupiter, Saturn, and Uranus, astronomers had two choices: either throw out Newton's equation or find the cause of the irregularities. They tried the second choice and found the planet, Neptune, having the exact location and mass that was predicted by Newton's equations.

Another is the explanation for the irregularity found in the movement of Jupiter's moons – the explanation, incidentally, that gave us the first real estimate for the speed of light. Astronomers observed that the moons were ahead of 'schedule' (as predicted by Newton) when Jupiter was closer to Earth and behind 'schedule' when Jupiter was farther from us. This discrepancy is explained by the speed of light: it takes light longer to reach Earth from farther away.

The full power of Newton's laws became apparent during the following two centuries after the science of Newtonian Mechanics was developed and its application to previously unrelated disciplines, such as thermodynamics and acoustics, was demonstrated.

Newton's laws can be used to explain even very complicated phenomena. Consider the following brain teaser: A fly and an express train move toward each other. The speed of the train is 100 kph, the speed of the fly is 10kph. When they collide, the speed of the fly has to drop from 10 kph to zero, before it starts accelerating backward. In that billionth of a second, the speed of the fly was zero. Since it was in contact with the train, the speed of the train also had to be zero for a billionth of a second. Correct?

The logic seems infallible, yet we know that it is nonsense. So, what is wrong with the logic I just described? I will give you the answer at the end of the next chapter, by which time you will have hopefully figured it out yourself.

Chapter IV – The Science of Mechanics

Newton's three laws, with their respective definitions, contain everything we need to solve almost all problems relating to motion: the orbits of the planets, comets, and moons; the design principles of rocket engines and spacecraft; structural requirements for bridges and skyscrapers; most of what we need for building clever machines to produce the conveniences of modern life.

Following Newton, many talented scientists fell on his system like wolves on a fresh kill: they developed it, formalized it, derived new concepts and new methods of calculation from it. They created the science of "Newtonian Mechanics".

Once people knew that all they had to do is find the formulas for different kinds of forces - and many of these formulae could be determined by doing simple experiments - Newton's second law and the mathematical tool of calculus gave them the means to determine how different material objects, subjected to these forces, would behave. However, scientists and engineers soon discovered that applying Newton's laws directly to different situations was sometimes mathematically difficult, so they set out to make the task easier by deriving some consequences of Newton's laws first.

They discovered new concepts such as momentum, energy, work, power, potential, angular momentum, and torque. They also discovered that the total value of some of these attributes never changes in physical interactions.

That is how they discovered the Laws of "Conservation of Mass", "Conservation of Momentum", "Conservation of Energy" and "Conservation of Angular Momentum". These conservation laws proved extremely useful because they helped to solve problems where applying Newton's laws directly would have been much more difficult.

For those who are curious and feel confident in handling high-school mathematics, I will deduce these conservation laws from Newton's Laws and also do the calculation for the rocket engine's escape velocity requirements and other practical examples in the last section of this chapter, called "The Next Level".

Conservation of Momentum

Most of you have played billiard or pool, or at least seen it played. I have always been fascinated by the 'science of collision' as manifested by the precise and predictable paths those smooth billiard balls follow when we shoot one against the other.

Billiards is, to a large degree, a mental game - a game of Newtonian physics. We expect certain behavior from the balls, for example traveling in straight lines and bouncing off each other at definite angles, and would be very surprised if a ball started rolling in the 'wrong' direction. The balls can't help but follow Newton's equations. It's in their nature, being part of the universe. The example of billiard balls has been used to great advantage in all areas of physics (thermodynamics, atomic physics) where we imagined matter being made up of countless little balls (molecules, atoms, etc.) and wanted to see what Newtonian physics would predict about the behavior of these systems. So, what is momentum and how is it conserved?

I hope you still remember Newton's original formulation of his Second and Third Laws:

Law II
"The alteration of motion is ever proportional to the motive force impressed; and is made in the direction of the right line in which that force is impressed"

Law III
"To every action there is always opposed an equal reaction: or the mutual actions of two bodies upon each other are always equal, and directed to contrary parts"

In the Second Law "motion", according to Newton's **Definition II),** means "arising from the velocity and quantity of matter conjunctly".

This is what physicists called the "momentum of a material body": its mass multiplied by its velocity.

We know from experience that stopping a rolling cart is harder if either the speed or the mass of the cart is greater. If either of these or both, is greater, we sometimes say: the cart has greater momentum. So, we see that the concept of momentum describes something important in nature.

What else can we say about the momentum that will help us solve complicated mechanical problems?

With simple math, we could deduce from Newton's Second and Third Laws the "**Conservation of Momentum**" law:

If there is no external force acting on a mechanical system (such as a billiards game) then the total momentum of the system (made up of the sum of the momentums of the individual components) remains the same after the interaction between the component parts.

This conservation law is very useful: scientists at NASA use it to calculate the maximum speed a rocket can reach in space, depending on the mass of the rocket, the mass of the fuel it carries, and the ejection speed of hot gases from the rocket engine (see the "Next Level" chapter).

The example of an elastic collision between two billiard balls is often used in nuclear physics. If two billiard balls collide, the velocities of the two billiard balls after the collision will be determined by the velocities of the two balls before the collision. Since the masses are the same, we assume that the sum of post-collision velocities will be the same as the sum of pre-collision velocities. We expect it to happen this way, but we don't always know that we count on the balls obeying the "Conservation of Momentum" consequence of Newton's second and third laws.

The "Conservation of Momentum" Law determines only the sum of the post-collision velocities of the two billiard balls. To find out the exact velocity of each, we will have to take into account another conservation law; that of kinetic energy.

Kinetic Energy; Work; Kinetic Energy Theorem

What other useful laws can we deduce from Newton's equations?

A concept with which we are already familiar is that of 'energy'. At least we have some feel for what it means. Ignoring all the other forms in which energy can manifest (heat, electric, nuclear, etc.) I will talk only about mechanical energy in this section. Let's start again with a few familiar examples:

Imagine that you are riding a bicycle on a perfectly smooth, horizontal road, accelerating from zero to 30 km/h in 10 seconds, using all your 'force' on the pedals. You would agree that it is hard work. What did you get in exchange for your hard work? Besides nearing your destination, you have gained speed. Your bicycle now has speed that can be converted into all kinds of useful work, like carrying you up a slope, without any further effort on your part.

Two concepts emerge from this simple example: "work" and "energy".

In physics, **Work** is defined as exerting force on a material body and making it move on some path.

(Kinetic) **Energy**, on the other hand, is defined as one-half of the mass of the object, multiplied by the square of its speed.

The concept of work is easy to grasp, we are intuitively familiar with it.

The definition of energy, as given above, seems a bit convoluted: We intuitively understand that the bigger the mass and the bigger the speed, the bigger the motional (kinetic) energy of the object should be. But why one-half of the mass? Why the square of the speed? The answer, as previously, comes from Newton's equations:

With simple math it can be deduced from Newton's second law that **the change in the kinetic energy of a material object equals the work performed on it** (as the terms were defined above). This seems to make sense because there is a direct exchange here: we put work in, we get energy out. This rule is called the "Kinetic Energy Theorem".

But why the square of the speed instead of just the speed?

Every driving instructor will tell us that the damage done to a car when it hits a tree does not merely double if we double our speed: it will quadruple. And if we triple our speed, the damage will be about nine times as severe. This means that the 'work' done on our car (many, many material particles moved some distances out of alignment by the application of considerable force on them) will increase with the square of our speed.

But why square, instead of cube, of the speed? I can give you two answers, neither of which will please you. One is that Nature has picked the 'square' without consulting us. The other is that if we mathematically integrate Newton's second law then both the one-half and the square pops out of the integration, as required by the rules of calculus (see "The Next Level" chapter).

Personally, I prefer the first answer: It has been like that since the Big Bang, and it is one of those things that we have to accept as given.

What good is this Law of Kinetic Energy for practical purposes?

Oh, it is a very useful tool indeed. For example, we can use it to calculate the escape velocity of a rocket-engine-driven spacecraft.

Imagine that a rocket is accelerated from the surface of Earth, vertically up to a certain velocity, when the engine runs out of fuel and stops. From that point on, the rocket decelerates (the rocket's speed steadily decreases), due to the gravitational force of Earth acting on it.

How large should the cut-out speed be so the rocket does not fall back to Earth? The escape velocity, by definition, is the speed the rocket needs to escape the gravitational attraction of Earth. Therefore, we can ignore any other masses in the universe, because their presence does not affect the gravitational attraction of Earth or the escape velocity determined by it. We know that if the rocket does not fall back to Earth, then it will keep going forever. If it did not keep going forever, then when its speed was reduced to zero, it would start accelerating back toward Earth, since the gravitational attraction of Earth is never zero, however small it gets with distance. Therefore, we assume that the velocity of the rocket is reduced to zero at an infinite distance from Earth. Of course. another mass such as the moon may capture it if the rocket flies too close, but it is irrelevant as far as the escape velocity is concerned. The concept of escape velocity only asks: what should the rocket's speed be not to fall back to Earth if it could go freely, without outside interference. We can calculate both the work and the energy change from Earth to infinite distance. According to the Kinetic Energy Theorem, the kinetic energy change (loss) after decelerating from the escape velocity to zero equals the work done on the rocket by the gravitational force. This will give us an equation for the escape velocity. When we solve this equation, by substituting infinity to the distance where the rocket's velocity reaches zero, we find that the escape velocity of the rocket is 1.1×10^4 m/s (11km/s), independent of the rocket's mass.

That is how simple rocket science is, once you know and understand Newton's laws! The same laws are being used today by real rocket scientists at NASA!

Small print disclaimer: they have to learn a few more details before they are allowed to design space ship engines.

Potential Energy, Conservation of Energy Principle

Imagine that you throw a ball up in the air with as much speed as you can manage. When you let go of the ball, it has a definite kinetic energy which is half its mass multiplied by the square of its speed. As the ball flies upward, its speed gradually decreases until it reaches zero at its maximum height. Objects with zero speed have zero kinetic energy.

So, what happened to it? Where did it go? We would expect to have gained something in exchange for the energy the ball has given up. We would be right, and yes, the ball has gained something. What it gained was "**Potential Energy**". It is easy to prove that in a gravitational force field any work done by moving an object from one height to another depends only on the difference in height and not on what trajectory we choose for the object.

This means that if we measure everything from sea level, then we can assign a number to any point in space and this number will be the work required to move an object from sea level to the height of that point. In Earth's gravitational field, the potential energy possessed by an object is proportional to the mass of the object and the height of its position above sea level.

Since, according to the Kinetic Energy Theory, the change in kinetic energy (half of the mass times the square of the speed) is *also* equal to the work done on the object, we can see that in Earth's gravitational field, the total mechanical energy of an object (kinetic plus potential) remains the same: as the ball flies upward, it loses kinetic energy but gains potential energy in such a way that the sum of the two will remain the same.

When the ball reaches its maximum height, all of its kinetic energy has been converted to potential energy (maximum height, zero speed). After that it will start falling down, reversing the process; gaining speed, and losing height, until it falls on the ground with a maximum speed (and kinetic energy) and zero height (potential energy).

If we do not consider any energy loss due to friction, air drafts, etc., the mysterious quality of the object, called **total mechanical energy** remains exactly the same.

Why?

Nobody knows.

The universe has obeyed this law ever since the Big Bang, to the delight of all physicists who use it to calculate solutions to problems that would otherwise be very difficult to solve.

The Law of Conservation of Angular Momentum

Of all the conservation laws in Mechanics, this one is my absolute favorite. The reason for this is the law's totally mystifying nature. We can sort of understand the conservation of matter, energy, and momentum, but rotational motion is not something we intuit.

Expressed in plain English, the law says: **if there is no outside torque affecting a material body, then its total angular momentum has to remain the same, no matter what the body does by itself.**

To understand it, we have to define torque and angular momentum.

We have a fairly good idea of what **torque** means, especially those of us who have ever used a torque wrench or tried to close a door pushing on it at different distances from the hinges. We know that the farther from the hinge we push on a door, the easier it is - the less force needed - to close because we are able to increase our torque by getting farther away from the axis of rotation.

Torque around a given axis is the rotating effect of the force applied to an object. Torque is obviously proportional to the force and the distance from the axis. More precisely, only that component of the force counts which is perpendicular to the line from the point of contact to the axis. If there is no perpendicular component (like pulling or pushing on the door edgewise in direct line to the axis) then there is no torque. The door doesn't close, either.

Angular momentum has to do with how hard it is to start or stop an object from rotating. From experience, we know that the force required to stop a rotating object depends on the mass distribution of the object around the axis and its speed of rotation. The more mass, the farther away from the axis, and the faster the object rotates, the harder is to stop it. Therefore, its angular momentum will be greater.

And, just as objects have inertia that keeps them moving in a straight line and at uniform speed if there is no force acting on them, we have the equivalent principle for rotating bodies. If there is no torque acting on a rotating body, it will keep rotating at the same speed*, as long as its mass distribution remains the same.*

This is a delightful complication unique to rotating motions, as compared to a straight line, uniform speed motions, due to the fact that the angular momentum not only depends on the mass and rotational speed but also on mass distribution! This leads to totally unexpected behavior of rotating bodies that are outside of our daily experience with moving bodies.

Examples are many, some of which I have already described: the pirouette performed by the ice skater; the speeding up and slowing down rotating motion of a high diver as he balls up or straightens out his body; the piano chair and the bicycle wheel combination I mentioned before.

In all of these examples, as the mass distribution changes, the rest of the system has to compensate, so the Law is obeyed.

This law can be deduced mathematically from Newton's Second and Third Laws (see "The Next Level" chapter), but there is an intuitive gap: just because force equals mass times acceleration (second law) and all internal forces cancel out (third law), I still can't 'feel' why an ice skater's body has to spin faster as she pulls her arms closer to her body.

We have many other aspects of Newtonian Mechanics that are fun to deal with, but this book is about basic principles, not applications. All the basic principles are contained in Newton's three laws and his gravitational formula; the rest was just an illustration of how powerful these laws are and how much is contained in how little.

Finally, as promised, here is the answer to the brain teaser I mentioned earlier:

Repeating the puzzle:

A fly and an express train move toward each other. The speed of the train is 100 kph, the speed of the fly is 10 kph. When they collide, the speed of the fly has to drop from 10 kph to zero, before it starts accelerating backward. In that billionth of a second, the speed of the fly was zero, and since it was in contact with the train, the speed of the train also had to be zero for a billionth of a second. Correct?

The precise answer, given by physics, is the rejection of the unstated (and maliciously smuggled in) assumption that both the fly and the train are rigid bodies. They are both composed of billions of molecules. Guess which of them has more! While individual train- and fly- molecules collide as quasi-rigid bodies and bounce back, the rest of the train molecules keep moving forward, easily shrugging off the slight elastic tug exerted on them via the lattice-like molecular structure of the metal. But, because of the law of 'conservation of energy' in physics, the train's speed will also drop an infinitesimal amount to account for the change in the kinetic energy of the fly molecules.

Range of influence

Newton's Laws and the system built upon them can be used in any area that involves forces and motions. For example, take thermodynamics, the science of heat. First, we see no connection to forces and motion: after all, a body can be hot or cold, and change from one to the other, without anything moving or visible forces acting on the body.

It took some imagination, but people started wondering if the heat had anything to do with motion after all. In 1798, while boring cannons for the Bavarian army, Benjamin Rumford noticed that the metal grew hot as the boring tool gouged them out, so they had to be cooled with water. Evidently, the heat had something to do with motion.

James Prescott Joule, in 1847, determined the conversion factor between mechanical work and heat: 41,800,000 ergs of work produces one calorie of heat. From Joule's discovery, it was only a logical step to assume that, according to the then still debated atomic theory: the heat was due to the random movements of atoms and molecules. The faster they moved, the greater was the temperature of the object.

Ludwig Boltzmann, in 1871, was the first to work out the "kinetic theory of gases" based on Newton's laws and the model of randomly moving and colliding particles. His theory was further developed by James Clerk Maxwell in 1871 and, as a result, we have a theory of heat and thermodynamic processes, based on Newton's laws of motion. Of course, this theory is not complete or entirely correct, because the behavior of small particles is subject to quantum mechanical laws, but the theory successfully explained many observed thermodynamic properties of gases.

Slowly, one by one, the laws governing all other phenomena were fully or partially deduced from Newton's equations: fluid dynamics, acoustics, friction, elasticity, etc. Nothing seemed to be outside its sphere of validity until the phenomena of electromagnetism and relativity were tackled at the end of the nineteenth and beginning of the twentieth century. Then all hell broke loose and wiped the complacent smiles off the faces of over-confident physicists all over the world.

Chapter V – Electricity and Magnetism

19th Century Science

With the discovery of electromagnetic phenomena, the Newtonian clockwork universe started to crumble. To the inexplicable *action at a distance* of gravitational attraction, now we had to add the same kind of attraction and repulsion between electrical charges and between magnetic poles.

Then it was found that moving electric charges created magnetic effects and moving magnets induced electric currents. Michael Faraday invented the concept of the 'field' and even though it turned out to be highly visual and intuitive, it was still only a hypothetical 'something', owning certain properties and having a definite structure, but no one actually knew what it was 'made of' and what it 'looked like'.

The progression of scientific thought from the mechanistic, rigid logic started by Newton, to the almost whimsical play with the esoteric concepts of Einstein's space and time was the most dramatic change science had ever experienced.

We had reached the limits of common-sense personal experience and embarked upon a journey that carried our minds beyond familiar boundaries. We could no longer simply describe what we saw because we could no longer see what we were talking about. Now we could describe reality only with mathematical equations or visual analogies.

Hidden in this drive to understand was our desire to grow beyond our limitations, to have at least a glimpse of what the universe was really like in regions of extreme speeds, forces, masses, and energies.

We have come full circle since Plato's cave. After centuries of rejecting Plato's mysticism in favor of reason, now we had to admit that the reality we humans can experience is a mere shadow of a much richer and much more complicated universe.

If we have a chance to look at an avalanche in progress, we can see that it starts slowly. A shift in the weight distribution of the mass of snow, a softening of rigidity and resistance, a fissure along the length of the field – and gradually it begins to move, to tumble, gathering speed as it thunders down the mountain.

So, it was with physics at the end of the 19th century. As more and more knowledge was amassed by busy scientists studying electricity and magnetism, more and more concepts emerged that caused a widening fissure in the body of Newtonian Mechanics. It was Albert Einstein who gave the final push to the avalanche in modern science. By his Special Theory of Relativity, he forced physicists to make the leap from the familiar Newtonian world to the frightening uncertainty of special and general relativity.

I remember how amused we were in high school when we learned about the famous quote attributed to Oliver Cromwell at the Battle of Dunbar, on September 3, 1650. A nervous Cromwell spent the night riding from regiment to regiment by torchlight on a small Scottish pony, telling his troops to: "Remember our battle-cry: the Lord of Hosts! Put your trust in God, my boys -- and keep your powder dry!"

Almost exactly a hundred years later, in 1752, Benjamin Franklin published, in *Poor Richard's Almanac*, his experimental results with Leyden jars, showing that they discharged electricity faster if a sharp needle was attached to them. Soon after, lightning rods appeared on church steeples.

These two examples show that all through human history, faith and knowledge were vying with each other for supremacy in man's mind and heart. This rivalry, however, was unnecessary, because they serve different purposes. We need some kind of faith to set goals and we need reason and knowledge to reach them. Human beings possess material bodies and live in a material world. Survival requires material sustenance, and we can be destroyed by material forces. No wonder that at the beginning of scientific investigation we wanted to understand the two most obvious phenomena affecting our survival: movement and forces.

Newton gave us a start with his Laws and his suggestion that we should find the laws of forces in nature. His gravitational formula gave us the first objective description of the most visible force affecting all of us: gravity. Other forces were known to men: elastic forces, forces of friction, centrifugal forces – all clearly visible, all strictly mechanical.

First Steps

Some forces, of which we were dimly aware, seemed mysterious, like the force of gravity itself: not requiring contact in order to operate. These were electric and magnetic attraction and repulsion. **Thales** in 585 B.C discovered that a certain kind of ore attracted iron. Thales also studied amber, which, when rubbed, attracts light objects.

Finally, in the 12th century, scientists started to take a serious new look at these forces. Our knowledge of them accumulated very slowly at first, then with increasing momentum over the centuries.

•   In **1180** the English **Alexander Neckam** described the direction-finding ability of lodestone, a naturally occurring magnetic ore. If it was allowed to move freely, for instance, attached to a cork floating on water, it always came to rest in a north-south direction. Very soon after this, European navigators started using magnetic compasses on the open seas.

•   In **1269** the French **Pelerin de Maricourt** described magnetic poles, stating that all magnets have two poles, north, and south; that opposite poles (north-south) attract, while identical poles (north-north or south-south) repel each other. He also discovered that the two poles cannot be separated by breaking the magnet in half. He also suggested mounting the compass needle on a pivot.

•   In **1600** the English **William Gilbert** published a book, *De Magnete,* describing his experiments with magnetism. He made a globe out of a large piece of lodestone, located its magnetic poles and demonstrated how a compass would point toward its "north pole", directly through the body of the globe (not tangentially to the surface), a *'magnetic dip'* that has been known since 1576. By this experiment, Gilbert demonstrated that the Earth itself is a huge magnet.

- In **1635** the English astronomer **Henry Gillebrand** demonstrated "magnetic declination", which meant that the magnetic north pole of Earth did not coincide with the true north pole (rotational axis point of Earth) and the magnetic north pole also changed its location over the decades.

- In **1660** a German physicist, **Otto von Guericke** made a globe out of sulfur and showed that when rubbed by hand while rotated on a shaft, it would build up an electric charge. The globe could be charged and discharged (producing a spark by grounding) any number of times.

- In **1706** the English physicist **Francis Hauksbee** replaced the sulfur with glass and succeeded in building up a much stronger charge than von Guericke had done.

- In **1729** the English physicist **Stephen Gray** discovered that electric charge could travel from one body to another. He hypothesized that electricity was some kind of fluid. Further investigations showed that this 'fluid' could travel more easily in some materials than in others. From this experiment, the concepts of *conductors* and *insulators* were deduced.

- In **1733** the French scientist **Cisternay du Fay** discovered that certain electrified bodies attract each other, while certain other kinds repel. From this, he deduced that there must be two kinds of electric fluid. He called one kind *vitreous* (Latin for glass) and the other kind *resinous* (from the resin rod he used). This was the first time that similarity was found between electricity and magnetism.

- In **1745** Dutch physicist **Pieter van Musschenbroek** filled a metal container with water, suspended it from an insulating thread, and built up an electric charge in the water by immersing a brass wire led from a friction-ball machine. Since he worked at the University of

Leyden in the Netherlands, his device was called the *Leyden jar* and was used thereafter for storing electricity. It also accidentally gave people nasty electric shocks.

- In **1747** American scientist and statesman **Benjamin Franklin** suggested that there was only one kind of electric 'fluid'. In access, the fluid produced one kind of electricity, which he called positive, and in deficiency, the fluid produced the other kind, which he called negative. In 1751 he performed his famous experiment with the kite in a lightning storm: leading the electric charge from the silk thread holding the kite into a Leyden jar. Thus, he proved that the charge obtained in this way was identical to that produced by friction.

- In **1777** French physicist **Charles Coulomb** invented the torsion balance. In 1785 he used it to demonstrate that the force of attraction or repulsion was proportional to the product of the charges and inversely proportional to the square of the distance between them. This fact is now called Coulomb's law: $F = k_e Q_1 Q_2 / r^2$ where k_e depends on the unit we choose for the electric charge. Coulomb established a similar law for the force of attraction or repulsion between point-like magnetic poles: $F = k_m p_1 p_2 / r^2$ where k_m depends on the unit we choose for the magnetic pole.

- In **1791** Italian professor of anatomy **Luigi Galvani** discovered that animal tissue (dissected frog legs) would twitch due to a spark from an electric machine. When the tissue was hung by brass hooks on an iron fence, it also twitched during a thunderstorm.

So far, all the experiments with electricity concerned static (unmoving) charges, due to the fact that no one yet was able to generate continuous electricity that could be led from its source in a steady flow. All they knew, before Volta invented his battery in 1800, was the charge and discharge of different bodies. Once the battery was invented, the study of electricity accelerated, and dramatic results were produced almost daily in physics laboratories.

One more thought it is difficult to appreciate these early results from the lap of the twenty-first century. Now we know 'everything' about electricity: when the battery runs low in our iPhone, we replace it, and when we want to watch our new TV, we plug it into the wall receptacle. What could be simpler?

Yet even in this modern age, most people who know that electricity is actually the current of electrons flowing through a wire, have no idea how long it would take one electron to get from the generating station into your television set. Most would be surprised to learn that it could take years, depending on your distance from the electric power plant, because electrons travel in the wire at the colossal speed of one meter per hour. This is called *drift velocity*. Yet, electricity starts flowing instantly through your TV set. We shall understand this later, after learning about electric fields.

The early scientists I described knew nothing about electrons, batteries, power plants, and wall receptacles. They had to start at the beginning with glass rods and silk scarves and lodestone. Until we understand this, we may be tempted to feel superior to those bungling amateurs who were trying to make sense of all that was visible at the time.

Alesandro Volta (1745 – 1827)

Volta was a curious, talented, ambitious man. One of nine children and no prodigy - he did not talk until four years of age - he had very limited education in a Jesuit college and then seminary. His interest in "natural philosophy", as physics was called at the time, set him on a course of self-study in various libraries. By the age of eighteen that turned him into an accomplished amateur, especially in the field of electricity.

This was not unusual during the "Enlightenment". What was unusual was his audacity to contact Nollet and Beccaria, the foremost experts in Italy. He sent them letters regarding his theories on electricity and magnetism. Since he had no credentials of any kind and no personal connections, it took him three years of stubborn and persistent trying until both scientists recognized his talent, and finally a steady and regular correspondence was established.

Since his family was not rich and he needed an income, he tried very hard to get a job in the Italian education system. Finally, in 1774, in a letter to Carlo di Firmian, the plenipotentiary Minister in Milan, he suggested that a new position of "superintendent" should be created in Italian public schools. He applied for "a Lectureship in Physics, Metaphysics or superintendent to public education in the Town of Como". For credentials, he submitted two papers on electricity that he had published in Latin, in 1769 and 1771. He succeeded and was appointed to the position he had created!

The administrative duties of a superintendent did not fully occupy his mind, so he set out to improve the Como educational system, in line with the Enlightenment ideas of "Innovation, reform, advancement of sciences and a happier culture". In recognition of these efforts, in 1775 he was appointed to the chair of experimental physics, in addition to his position of superintendent.

Volta had made it. He had a steady job, the status of an expert in his chosen field, an opportunity to advance as far as his abilities permitted. In that same year, he built his first successful electrical machine, the *Electrophorus*, and published its abstract in a Milanese journal. The electrophorus made Volta both locally and internationally famous.

The device built up an electric charge by first separating positive and negative charges in a metal plate, held by an insulating handle, and then grounding the negative charge from the top surface of the plate; thus the whole plate turned positive. An unlimited number of sparks could be created by touching the positive plate. This was a major improvement over the Leyden jar, which could be discharged only once.

As it turned out, the electrophorus was only a preamble to Volta's fame. His real significance, revolutionizing the science of electromagnetism, is based on the invention of the Voltaic pile, or electric battery. As soon as he read Galvani's report on his experiments with frog legs, he repeated the experiment himself.

In the ensuing correspondence, Galvani, a professor of anatomy, insisted that the source of the observed electricity was the body of the frog itself, while Volta the physicist was convinced that the source was to be found in the two different metals (brass hooks and iron fence) touching each other. The disagreement between Galvani and Volta, and their respective followers, developed into a feud that lasted beyond Galvani's death until Volta announced his first electric battery that produced continuous electricity by using silver and zinc and no frog legs or any other animal tissue.

At the time, no instrument existed that could measure very small electric charges; therefore, physicists used frog legs to detect them. It was cumbersome, to say the least. Volta developed the *Condensatore* which was very similar to his electrophorus, but its function was to show the presence of a very small electric charge, rather than create it.

The rest is well-known history. What is not widely known, however, is that in order to refute one of Galvani's arguments, Volta spent quite some time researching all that was known about the electric eel, also called torpedo fish. Among other things, he read an article published by the English chemist William Nicholson regarding his idea of constructing an electro-mechanical device that could duplicate the eel's ability to give continuous shocks. Nicholson suggested using Volta's electrophorus as the basic cell in the device.

This article gave Volta the idea that led to the invention of the electric battery. He set out to create a device that could produce a continuous flow of electricity, just as the electric eel seemed to be able to do. However, he greatly simplified Nicholson's idea: he connected the cells in a serial fashion rather than the parallel suggested by Nicholson. In the first Voltaic pile, a series of cups containing dilute sulphuric acid were connected by alternating zinc and silver conductors. When the two ends (zinc on one side and silver on the other) of the 'crown of cups' were connected by a conductor, a continuous flow of electric charge could be observed.

Later he simplified the arrangement and made it more portable by eliminating the cups. One main appeal of the device was the extreme simplicity of its construction. It could be precisely described with as few as 131 words:

"A number of pieces of zinc, each of the size of a half-crown, were prepared, and an equal number of pieces of card cut in the same form; a piece of zinc was then laid upon the table, and upon it, a half-crown; upon this was placed a piece of card moistened with water; upon the card was laid another piece of zinc, upon that another half-crown, then a wet card, and so alternately till more than forty pieces of each had been placed upon each other; a person then, having his hand well wetted, touched the piece of zinc at the bottom with one hand, and the half-crown at the top with the other: he felt a strong shock, which was repeated as often as the contact was renewed."

The invention created a sensation all over Europe. Even Napoleon invited Volta in 1801 to demonstrate his discovery. Isn't it curious? The great conquering hero, emperor of France, was fascinated by electricity!

Physicists, chemists, and other scientists immediately set out to duplicate Volta's experiments to build larger and larger electric batteries, to produce stronger and stronger electric currents.

Almost overnight electrolysis was invented, followed by the invention of electroplating, and many other breakthroughs in applied chemistry.

The greatest consequence of the now available steady electric current was the birth of electromagnetism twenty years later, by Oersted's discovery of the effect of electric current on the magnetic compass. Those twenty years were not exactly wasted by scientists: the Voltaic battery opened a number of paths for research and application and there was no shortage of curious minds to follow them up. Physicists acted like children with a new toy.

•   In 1802 English chemist Humphry Davy constructed a pile with sixty pairs of zinc and copper plates, each 6 inches square. This battery could melt iron wires up to 1/10 inch thick.

•   In 1807 English instrument maker William Pepys constructed an even stronger battery of 274 plates. Using this battery, Davy could decompose alkalis, extracting the new elements potassium and sodium.

- In 1808 Pepys constructed another, even more powerful battery of 2000 pairs of plates of zinc and copper that could produce spectacular electric arcs and which Davy used to extract the additional new elements barium, strontium, calcium, and magnesium.

- In 1809, English naturalist John G. Children built a very large battery with twenty pairs of copper and zinc plates, each plate 6 feet long and 2 feet, 8 inches wide.

For these ever more powerful batteries, ever more new applications were found.

- As we have seen, chemists started decomposing different compounds and extracting previously unknown elements. Chemical decomposition is the separation of a complex chemical compound into elements or simpler compounds. Electrolysis is the decomposition of non-metallic conductors with electricity. For example, in 1800 William Nicholson ran an electric current through water that resulted in decomposing it into oxygen and hydrogen.

- A new way of making steel was discovered when Pepys, in 1815, melted a short section of iron wire and diamond dust together, thereby directly carburizing the iron.

- Italian chemist Luigi Brugnatelli, Volta's colleague, experimented with electroplating. This is a process of electrolyzing a metal in solution so that a very thin layer adheres to the surface of a different metal. He succeeded in coating two large silver medals with gold.

- Military applications were not far behind. During the occupation of Paris, in 1815, the allied armies used an electric fuse to produce a spark to ignite gunpowder at a distance.

- In 1809, German physician Samuel Soemmering invented the 'electrochemical telegraph', using 35 wires connected to 35 gold rods set into glass tubes containing acidulated water. When any one of the wires was connected to a battery, the corresponding glass tube, at a distance of 1000 feet, would produce gas bubbles. This very clumsy and complicated way to send messages from one end to the other was, nevertheless, the first time it could be done at all.

In addition to building larger and larger batteries and finding practical applications, scientists performed many experiments in order to understand the phenomena themselves. For example:

- In 1804 Italian physicist Giovanni Aldini (Galvani's nephew) tried to determine the speed of an electric current. All he could conclude was "astonishing rapidity". He had no idea that the actual speed could circumnavigate the Earth seven times in one second!

- American chemistry professor Robert Hare determined that the intensity of the electric current increased with the number of plates used, while the quantity of electric flow depended on the size of the exposed plates.

As a prelude to Oersted's discovery, several attempts were made to find a relationship between magnetism and electricity.

- In 1801, French chemist Nicolas Gautherot observed that two wires, each connected to the two ends of an electric battery, would tend to adhere to each other when they were brought near.

- In 1806, French physicist C.J. Lehot repeated the experiment with the same result. The phenomenon was observed by two famous French physicists, Biot and Laplace, as well.

- In 1805 French scientist Charles-Bernard Desormes set a battery on a float resting on water to find out whether the battery would behave like a magnet and orient itself with the Earth's magnetic field. No such orientation was observed.

Hans Christian Oersted (1777 - 1851)

Oersted is not widely known by the general public, even though he performed the first crucial experiment connecting electricity and magnetism. He never followed it up, but announced the breakthrough to the world and let others study it further. That reticence may explain his lack of fame. His career path took him from his father's apothecary shop in the small Danish town of Rudkoebing to the University of Copenhagen, where he obtained the degree of Doctor of Philosophy (not Physics or Chemistry) in 1799. In 1806 he became Professor of Physics at the same university and, intrigued by the Voltaic battery, he started looking into a possible relationship between electricity and magnetism.

Oersted was an ardent disciple of Immanuel Kant, a German philosopher whose most famous work, *Critique of Pure Reason*, was published in 1781. In this book, Kant tried to find unity between such apparently distinct phenomena as matter and spirit (science and religion). Kant's solution to this duality was a definition of matter as nothing but the manifestation of attractive and repulsive forces. While in Germany, Oerstead published, in German, his views of the chemical laws of nature:

"One has always been tempted to compare the magnetic forces with the electrical forces. The great resemblance between electrical and magnetic attractions and repulsions and the similarity of their laws necessarily would bring about this comparison. ….An attempt should be made to see if electricity, in its most latent stage, has any action on the magnet as such "

Kant's influence is clearly visible and, as we see, Oersted was on his way to his historic discovery of 1820. The first step was obvious: place a magnetic needle near a wire carrying an electric current and see if there is a reaction. Oersted - and many others - did that without any result. Unfortunately, they laid the conducting wire across the compass needle, instead of parallel with it.

What an irony!

One can just imagine all these bright people, repeating this experiment exactly the same way, without even accidentally placing magnet and wire parallel with each other, even for a moment while moving them about. Some may have actually done so and failed to notice the reaction of the needle. Another possible cause for the delay in discovery could have been the use of batteries too weak to produce a noticeable effect. They all consciously or unconsciously assumed that the force, if there was any, would act in the direction of the flow of electricity, and this parallel force should deflect a magnet perpendicular to it. No one expected the force to be perpendicular to the wire, *and* perpendicular to the line connecting wire and magnet. No force in nature acting in this strange way had ever been observed!

I know this is very confusing, but I can make it simpler with the help of two pencils. Hold up one pencil in your hand and point it away from you. This pencil will represent the flow of electric current in the wire. If you expect the force (if any) to be parallel with the wire, then with your other hand hold another pencil at right angles to, and above, the first one, expecting it to be rotated by the force. It won't rotate because it is already in its equilibrium position. However, if you expect the force to be perpendicular to the wire, then hold the other pencil parallel with the first one, expecting it to be rotated by the force. This time your second pencil will rotate and come into equilibrium in the perpendicular position because it matches the direction of force at that point.

Oersted finally discovered this effect in April 1820, during a lecture to his students, but *waited three months* before he found free time to pursue the matter. One wonders whether he was aware of the dramatic importance of his results for the scientific community of Europe. In a competitive world where scientists vie over priority and often risk serious embarrassment by rushing into print lest one of their colleagues beat them to the credit, it is almost unbelievable that Oersted waited that long with further experiments. Finally, on July 21 1820, Oersted published a four-page tract, in Latin, under the title: *Experiments on the Effect of a Current of Electricity*, in which he drew the following conclusions:

1.   The electric conflict acts on magnetic poles.
2.   The electric conflict is not confined within the conductor but also acts in the vicinity of the conductor.
3.   The electric conflict forms a vortex around the wire.
4.   If the direction of electric flow is reversed in the wire, then the force acting on the needle is reversed as well.

The reaction to this announcement, all over Europe, was spectacular! Ranging from enthusiasm to plain disbelief, every prominent physicist set out to duplicate Oersted's experiment… which was not too difficult. They merely had to move a wire next to a compass, in a parallel position with the needle, and turn on the current. No great discovery in the history of physics has ever been easier to test!

Oersted left further investigations of his discovery to others. He devoted the rest of his life to improving science education in Denmark and traveled and lectured widely in Europe. As a final recognition of his contribution to physics, the unit of magnetic field strength was officially named the 'oersted' in his honor in 1934.

An amusing coincidence. A collection of Oersted's essays was published in London in 1852, under the title *The Soul of Nature*. I was unaware of this when I chose the original title of this present book: "*Physics: The Soul of Nature*". After some soul-searching (pun intended) I decided to change it.

André-Marie Ampère (1775 – 1836)

Ampere did not have a happy start in life. In 1792, when he was 17 years old, his father was guillotined by the Republicans during the French Revolution. This tragic event threw him into a deep depression and withdrawal for two years. Shortly after his recovery, he fell in love with a young lady who, after three years of intensive courting, agreed to marry him in 1799. The marriage was happy but lasted only four years, during much of which they had to live apart due to Ampère's teaching duties in Bourg. Soon after Ampère managed to secure employment close to home in Lyon, his wife died of a mysterious illness, pushing him back into despair.

To escape the sad memories of Lyon, he moved to Paris in 1804, where he had no friends and never felt at home. Probably due to this loneliness, he married again in 1806, with disastrous consequences. His second wife did not wish to have children and when she got pregnant anyway, she threw Ampère out of their house and refused to have anything to do with him from then on. He learned of his daughter's birth from a porter. His wife rejected the child, who was put in foster care until Ampère was granted a separation decree and legal custody in 1808. His mother, who joined him in Paris to help look after his family died two years later, adding to his burdens and his depression. In view of all these hardships, it is almost a miracle that Ampère accomplished as much in science as he did.

Ampère had no formal education; he was tutored at home and had almost total freedom to study whatever interested him. He had an incredible mind, both in the power of concentration and in memory: he read through the entire multi-volume *Encyclopédie* in alphabetical order and could recall, years later, whole articles on such esoteric subjects as heraldry.

Since his family lost most of its possessions in the revolution, Ampère's income from his inheritance was very modest – he had to work for a living. He secured a teaching position in 1802, as Professor of Mathematics and Chemistry in Bourg, then in 1804 as *Répétiteur* (tutor) of mathematics at the École Politechnique in Paris, a school for civil and military engineers. Finally, in 1824, he was appointed Professor of Experimental Physics at the Collège de France in Paris.

Ampere was the stereotypically absent-minded professor. Stories about him include writing equations on the back of a carriage and watching it drive off into traffic when he was halfway through; throwing his watch in the river and putting a stone in his pocket; once he forgot an invitation to dine with Emperor Napoleon.

Ampère's interests and accomplishments were primarily in mathematics and chemistry – he never seriously attempted to work in physics until after Oersted's discovery in 1820, when he was 45 years old. It is almost unheard-of in the history of physics for someone to start a productive career that late in life. But, in spite of failing health, he produced spectacular results between 1820 and 1826. He created and named the science of *Electrodynamics* that became the foundation for all subsequent theoretical physics on the subject.

The whirlwind of activities following the announcement of Oersted's discovery at the Académie des Sciences on Sept 11, 1820, is best illustrated by the timeline provided in James Hofmann's biography: *André-Marie Ampère*: Highlights for the first three-month period:

18 September	Ampère demonstrates the tangential orientation of a magnetic needle by an electric current when terrestrial magnetism is neutralized.
25 September	Ampère demonstrates for the Académie that conducting spirals attract and repel each other and respond to bar magnets in an analogy to magnetic poles
9 October	At the Académie Ampère demonstrates electrodynamic forces between linear conducting wires
30 October	Ampère demonstrates the action of terrestrial magnetism on a suspended current loop. Biot and Savart make their first report to the Académie on their quantitative measurements
6 November	Ampère presents his addition law for electrodynamic forces and uses it to interpret the action of helices

| 4 December | Ampère presents his symmetry principle and uses it together with the addition law to derive the angular factor in his electrodynamic force law. |
| 18 December | Biot and Savart report to the Académie on their second set of measurements. |

From this timeline, we can see that two scientists responded to Oersted's discovery with immediate passion, both French: Jean Baptiste Biot (1774-1862) and André-Marie Ampère. Both performed numerous practical experiments on the magnetic effect of electric currents, and both developed hypothetical models for the cause and mechanism of the phenomenon. Both based mathematical calculations and deductions on their hypotheses and both claimed to have derived their laws from their models by mathematical reasoning.

Their main laws describing the magnetic force generated by the electrical currents were in different forms but, for the case of closed circuits, they described the same magnetic field as the function of current intensity and distance from the current.

There was only one difference: while Ampère's hypothesis was correct and his mathematics rigorous, Biot's assumptions were dead wrong, and his math was flawed. Yet, every college or university physics book today teaches the "Biot-Savart law", describing the magnetic field created by an electric current element. Luckily, Ampère could not foresee this 'injustice' at the time.

Ampère and Biot did not like each other. Much of their work was developed in competition, which was intense, polite, taking on tones of sarcasm and pettiness from time to time. They were both members of the Académie; they faced each other during the weekly meetings, like lawyers in a trial, referring to their "learned colleagues" in respectful voices.

Biot's contribution to the study of electromagnetic phenomena was small, compared to Ampère's. When Oersted announced his discovery of electric currents affecting magnetic needles, Biot took an immediate interest in the phenomenon and recruited a protégé, Felix Savart, to help him make the measurements.

They assumed that:

1.   Magnetism was due to the presence of two different magnetic fluids (*astral* and *boreal*) in both the magnetized needle and the conducting wire that was somehow magnetized by the current.

2.   The force observed on the entire magnetic needle was the same as the force on the assumed magnetic fluid particles inside the needle.

3.   The total observed force was due to the sum of all the forces exerted on the needle by very tiny wire segments

4.   The 1/r dependency of force on the distance was a consequence of a $1/r^2$ dependency of forces, exerted by the individual wire segments.

5.   The forces from individual wire segments were proportional to the sine of the angles between the wire and the line connecting the segment to the magnetic needle.

     Some of these assumptions (1 and 2) turned out to be totally false. The rest were only hypothetical and not mathematically proven, contrary to the claim by Biot and Savart. Nevertheless, the law we know today as "Biot-Savart" law states that the magnitude of the force exerted on a point-like magnetic pole of unit strength at point 'P', due to a tiny wire segment is:

$$dF(P) = K\, I\, ds\, \sin \zeta\, /\, r^2$$

where 'K' is a factor dependant on the units we chose for the electric charge and the magnetic pole strength; 'I' is the strength of the electric current; 'ds' is the length of the wire segment; 'ζ' (pronounced 'zeta') is the angle between the wire segment and the line to 'P' and 'r' is the distance of the wire segment from 'P'. In this form, the Biot Savart law is correct (without the incorrect interpretation) and can be used to calculate the magnetic force exerted by conducting wires of different shapes and sizes. For example, by using the Biot-Savart law, we can calculate that the magnetic force exerted on a magnetic pole of unit strength, by a circular wire, at the center of the circle equals:

$$F = K\, 2\pi\, I\, /\, a$$

Where 'I' is the strength of the current and 'a' is the radius of the circle.

Contrary to Biot and Savart, Ampère did not believe in 'magnetic fluids'. For him, all magnetic phenomena were due to electric currents. By attributing the magnetic properties of both Earth and magnetized metals (lodestone, steel, etc) to electric currents, he tried to unify different forces under one phenomenon (electricity), thereby eliminating the superfluous assumption of magnetic fluids. In this, he was absolutely right. Maxwell recognized this accomplishment when he describes in Ampère's work as: "one of the most brilliant achievements in science. The whole, theory and experiment, seems as if it had leaped, full-grown and full-armed, from the brain of the 'Newton of electricity'. It is perfect in form and unassailable in accuracy"

No, Ampère's theory did not leap "full-grown and full-armed" from his head. He worked very hard to develop it, through careful experimentation. When Oersted announced his discovery, Ampère's reaction to it was enthusiastic, to say the least. In quick succession, he produced some impressive experimental results:

- He invented an 'astatic magnetic needle' the rotation of which could be made perpendicular to the magnetic force exerted by the Earth, thereby eliminating the influence of Earth and isolating the phenomenon he studied.

- He demonstrated that there was a magnetic attraction and repulsion not only between current and magnet but also between two electric currents.

- He showed that a freely rotating helix of wire carrying electric current behaved exactly as a bar magnet would: reacting to both Earth and other bar magnets.

- He demonstrated the magnetic action between current elements (both parallel and at an arbitrary angle) and started deriving his force-law dependant on the strength of the current, distance of the current elements, and their mutual (three-dimensional) orientation.

- He invented a completely new kind of experimental method that is known today as the "null method". To the exasperation of his contemporaries, these experiments, if successful, resulted in 'nothing happening' - no movement, no action, no new data - by which Ampère proved that two forces, one known, another under consideration, were equal.

While Ampère produced these startling experimental results, he was steadily building a mathematical theory of magnetism, based on his initial assumptions, using very sophisticated tools. His decades-long experience and passion for mathematics proved very handy indeed.

At the end of this intense period, he published the collected results in his famous paper: "*Theory of Electrodynamic phenomena, based on deduction and experiment*" in 1826. This paper contained his **'force law'** between two infinitesimal current elements, and then, by applying this law to closed-circuit wires and solenoids, he deduced the well-known experimental results obtained both by Biot and Coulomb.

Ampère's force law states that the force acting between two infinitesimal current elements ds and ds' at distance 'r' from each other depends on the currents i and i' flowing through them and on three angles of their mutual 3-dimensional orientation according to the following formula:

$$df = i\ i'\ ds\ ds'/r^2\ (\sin\alpha\sin\beta\cos\chi\ -\ \tfrac{1}{2}\cos\alpha\cos\beta\ )$$

In the special case when the two current segments are parallel (α=90; β=90; χ=0) the force law takes the simple form of:

$$df = i\ i'\ ds\ ds'/r^2$$

from which we can calculate the force attracting or repelling two very long parallel wires as:

$$F = k_A\ I_1\ I_2\ L/r$$

Where k_A depends on our choice of unit current, I_1 and I_2 are the currents flowing through the two wires and 'r' is their distance from each other.

Michael Faraday (1791 – 1867)

Faraday was one of those rare individuals who are brilliant, honest, kind, ambitious, and very, very lucky. One shudders to think how many potential geniuses have been destroyed by bad luck and rotten circumstances. It warms the heart to know that, once in a while, the odds are beaten, and a miracle happens.

One of ten children of a blacksmith, young Michael had few options in life. At age fourteen, having learned to read and write, he was apprenticed to a bookbinder. That was the first bit of luck. Not only did he have access to books but, quite uncharacteristically for a tradesman, Master Riebau took an interest in, and encouraged, his apprentice. Faraday read everything that came through the shop, from the Arabian Nights to the Encyclopaedia Britannica. In fact, it was an article on electricity that first stirred his interest in science.

His passionate quest for self-improvement received help and direction from a book written by Dr. Isaac Watts, a clergyman of the eighteenth century, titled: *The Improvement of the Mind* - for "persons of younger years who needed a guide through the maze of learning".

Dr. Watts' book contained practical advice that Faraday immediately implemented in his own life, such as:

- keeping a notebook to record his thoughts and ideas
- attendance of lectures
- exchange of letters with like-minded persons
- joining discussion groups
- being observant
- being cautious about jumping to conclusions

Faraday followed this advice with almost religious zeal and soon found himself profiting from it. This barely educated young man soon built up an impressive knowledge of chemistry and electricity, an endeavor greatly helped by his master's permission to use the shop after hours to perform simple chemical experiments.

Moreover, Mr. Riebau gave a decisive boost to young Faraday's career when he showed one of Faraday's beautifully bound scientific notebooks to a client, who in turn showed it to his father, who was so impressed that he gave Faraday an admission ticket to a lecture by Sir Humphry Davy, England's foremost chemist, at the Royal Institution. Faraday attended four lectures by Davy and made copious notes of everything he saw and heard.

The apprenticeship was coming to an end by the fall of 1812 and Faraday was desperate to escape the drudgery of a bookbinder's life. He approached Davy, presenting the handsomely bound volume of notes he had taken at Davy's lectures, with drawings of the experiments, and asking for any position in his lab. Davy was impressed, of course, and, after he lost his assistant, he offered the position to Faraday.

This was the second piece of luck for Faraday. Better still, the job also included a trip across Europe with Davy - to France, Germany, and Italy - from the fall of 1813 to the spring of 1815. Considering that Faraday had never before been more than 12 miles from London, this was a great boost to his career. He met some of the prominent scientists of the day, and he learned French and Italian well enough to read scientific publications in those languages.

We know the rest: Faraday, without any formal education in science, became a world-class scientist, inventor, and major contributor to the development of Electromagnetism which, in turn, made the revolution in technology on which our modern lives completely depend. The story of Faraday illustrates how talent, determination, and exceptional luck can overcome almost all obstacles.

Before describing Faraday's contribution to physics, one more event in his personal life is worth mentioning because it was the final piece of luck Faraday needed to have the quiet, happy life prerequisite to his great accomplishments. In 1821 he married Sarah Barnard, who became his lifelong companion, his only true love and a lady who, when asked why she did not study chemistry, replied: "Already it is so absorbing and exciting to him that it often deprives him of his sleep and I am quite content to be the pillow of his mind".

Michael Faraday had it made - he had a career, a respectable position as a chemist with the Royal Institution, and a happy family life. He was thirty years old and it was time to start shaking the world.

Like most scientists in Europe, Faraday was very interested in the discovery of electromagnetic effects Oersted published in 1820. He used voltaic piles regularly in his experiments with chemical substances and was curious about the nature of electricity and magnetism. Influenced by Immanuel Kant's suggestion of an underlying unity of all forces of nature, the link between these two greatly intrigued him. He started serious work on electromagnetism in 1821 when he received a request by the *Annals of Philosophy* for an article on the history of this new branch of science.

Faraday's approach was typical. He repeated all the major experiments himself, making sure that he knew *exactly* what was happening. Then he read all the theories put forward by different scientists in the field. Out of the confusing jumble of assumptions and suggestions, he picked only Ampére's *Theory* as worth studying and analyzing in detail. Even though he did not understand much of the sophisticated mathematics used by Ampére, he was impressed by the simplicity and elegance of the theory and the seeming agreement between its predictions and the experimental results.

The difference between Ampére's and Faraday's approaches was obvious from the outset. Ampére was primarily a mathematician and theoretical physicist. He made assumptions based on philosophical arguments, developed a model based on these assumptions, and performed mathematical deductions from this model. He designed experiments for the sole purpose of proving his theory, then fine-tuned his theory according to the results. He did not do physics out of curiosity about nature and paid no attention to unexpected new phenomena unless they had a direct bearing on the theory he wanted to prove.

In contrast, Faraday did not know much mathematics and had no preconceived ideas but was full of curiosity and highly observant of everything happening around him.

In Faraday's view, it was important to make a very clear distinction between experimental results and theoretical assumptions. As he wrote: "There are many arguments in favor of the materiality of electricity, and but few against it; but still it is only a supposition; and it will be as well to remember, while pursuing the subject of electromagnetism, that we have no proof of the materiality of electricity, or of the existence of any current through the wire….Whatever be the cause which is active within the connecting wire, whether it be the passage of matter through it, or the induction of a particular state of its parts, it produces certain very extraordinary effects".

Before long, Faraday produced surprising new results not yet experienced by anyone. The first of these happened in September, 1821 and is now called "Electromagnetic Rotation". It was, in effect, the first **electric motor** ever created. After analyzing the forces that acted on a magnetic needle in the vicinity of the current-carrying wire, Faraday realized that the force should move a magnetic needle in a circle around the wire.

The experiment he set up was most ingenious. He stuck a magnet upright in the middle of a container filled with mercury. He suspended a wire from a stand above the magnet so that the upper end could freely pivot, and the loose lower dangled into the mercury. Then he connected one pole of a battery to the top end of the wire, and the other to the pool of mercury, forming a closed electric circuit.

The current created a magnetic field around the wire and this field, interacting with the magnet, started rotating the lower end of the suspended wire around the magnet, swimming through the pool of mercury.

This was the first time in history that the mysterious energy of the electromagnetic field was converted into the mechanical energy of motion.

Faraday deeply believed in the unity of all nature's forces and tried to visualize Kantian space filled with a mysterious force that permeates everything. One very convincing way he found to illustrate this force was to sprinkle iron filings on a board over a magnet. When he gently vibrated the board, the particles arranged themselves in a pattern that indicated the direction of the magnetic force acting around the magnet. He designed another experiment to illustrate how a freely moving magnetic pole would follow these lines of force even inside a magnet. Quoting from Faraday's own description of the experiment:

"A helix of silked copper wire was made round a glass tube…A magnetic needle…was floated with cork, so as to move about in water with the slightest impulse. The helix connected with the apparatus and put into the water in which the needle lay,…the needle entered the glass tube, but did not stop just within side in the neighborhood of this pole … of the helix, but passed up the tube, drawing the whole needle in and went to the opposite pole of the helix"

A brilliant experiment, showing very clearly what forces look like inside a magnet, represented by a current-carrying helix of wire. The only difference between a magnet and a current-bearing helix is in the location of the poles: in a real magnet, the poles are always a small distance from the ends, inside the magnet, while in a helix the poles appear at the very ends of the helix.

This and similar experiments led Faraday to conceptualize the electric and magnetic 'force fields'. His idea was in direct opposition with Ampére's Newtonian idea of attraction and repulsion between current segments of electricity. For Faraday, the circular force surrounding the wire was of a fundamental nature that could not be reduced to direct "action at a distance" forces between points in space.

Yet, Ampére continued to produce new startling experimental discoveries that were predicted by his theory. In 1821 December he announced the successful rotation of a magnet around its own axis, something Faraday had been unable to achieve. Obviously, both Ampére and Faraday had parts of the truth that needed to be integrated into one coherent theory.

Before such a theory could be developed, more experiments were to be conducted, more effects to be discovered, more analysis to be performed. The next major step took place 10 years later when Faraday discovered Electromagnetic Induction. The story of this discovery is an intriguing and amusing demonstration of how everything is truly connected; how seemingly unrelated phenomena have a role in major breakthroughs in science.

His first clue was a new electro-magnetic effect discovered by François Arago in France and Peter Barlow in England. Arago's experiment was simpler and so it is known today as "Arago's wheel".

A magnetic needle was suspended over a copper disk. When Arago rotated the disk, the needle above it started rotating as well. When he cut slits radially in the disk, the needle stopped rotating. No electric current was involved, the only movement of a non-magnetic disk, yet a magnetic effect was produced.

Trying to understand this mystery, Faraday started thinking about the mechanism of electromagnetic action. He did not believe in the electric two-fluid theory Ampére was using - instead, he tried to explain the magnetic phenomena (rotation, Arago's wheel) by the Kantian forces. But how could electrical forces be transmitted, without some kind of material, such as electrical fluids, moving through the wire? Faraday needed an idea about what form transmission could take.

This was to be the second clue, and he found it in the work of Augustin Fresnel who published his *Elementary view of the Undulatory Theory of Light* in a series of articles from 1827 to 1829. At the time, light was considered to be one of the 'imponderable fluids'. To Faraday's great delight, according to Fresnel's theory, light could be transmitted by wave motion alone, without the assumption of fluids of any kind. Here was the example he was looking for.

The third clue involved sound. He witnessed several demonstrations of the *kaleidophone* at the Royal Institution in 1828. In the experiment, sand was thrown upon two adjacent glass or steel plates and some pattern was created in the sand on one of them by rubbing a violin bow against the edge of the plate. A surprising effect was observed on the second plate: the sand on it also assumed the same pattern as on the first one, without any physical contact whatsoever. Evidently, forces could be transmitted through the medium of air by mere vibration.

These two examples prompted Faraday to consider whether electricity could also be undulatory in nature. To protect his priority, Faraday deposited a sealed note in the Royal Society's safe, stating that: "experiments, led me to believe that magnetic action is progressive and requires time…I am inclined to compare the diffusion of magnetic forces from a magnetic pole to the vibrations upon the surface of disturbed water, or those of air in the phenomena of sound; i.e. I am inclined to think the vibratory theory will apply to these phenomena, as it does to sound and most probably to light. By analogy, I think it may possibly apply to the phenomena of induction of electricity of tension also".

This idea was reinforced by another experimental result involving electromagnets that the American Joseph Henry was working on at the time. Apparently, the polarity of an electromagnet could be reversed almost instantaneously by reversing the direction of the current. This was in contradiction with contemporary theory - based on the assumption of the two electric fluids - which predicted that reversal would take a considerable time. On the other hand, if not fluids but waves were involved in the transmission of electricity, it could happen much faster.

All Faraday needed to do was fit the elements of the puzzle together. In August 1831, he performed his final, epoch-making experiment that became the first **electrical transformer** in history. Faraday discovered **electromagnetic induction,** which could create an electrical current in a remote wire by turning on (and off) electrical currents in another wire. Ten years of research culminated in the report:

"Have had an iron ring made…wound many coils of copper wire round one half…will call this side of the ring A. … on the other side but separated by an interval was wound wire…this side call B. …connected the ends… on A side with battery; immediately sensible effect on the needle [of galvanometer on 'B' side]. It oscillated and settled at last in original position. On braking connection on A side with Battery again a disturbance of needle. … effect due to a wave of electricity caused at moments of breaking and completing contacts at A side….Wave apparently very short and sudden"

Once Faraday realized that it was the change (starting and breaking the current) that caused the induction, he required one more step to create the first **electric generator:** a device to produce a continuous flow of electricity, not by chemical processes inside a battery, but by transforming mechanical motion.

First, he tried to shove a bar magnet inside a coil of wire and noticed that every time, while he was moving the bar magnet into or out of the coil, electric current flowed through the wire. Next, he rotated a copper disk between the poles of a U-shaped magnet and placed contacts on the edge of the wheel, leading them to a galvanometer. The experiment was a success. In his paper, *Experimental Researches in Electric*ity Faraday announced:

"Here, therefore, was demonstrated the production of a permanent current of electricity by ordinary magnets"

Every biographer's favorite quote concerns the visit by Sir Robert Peel, Prime Minister of England, who asked Faraday of what use was this new device. Faraday's reply: "I know not, but I wager that one day your government will tax it".

Faraday explained the phenomenon of electromagnetic induction, using his idea of magnetic lines of force that make up a magnetic field surrounding a current-carrying wire. As the current starts in the wire, the field builds up around it, and the changing intensity of the magnetic field induces the electric current in the other wire. When the contact is broken, the field starts collapsing and, again, the changing intensity of the field induces an electric current, flowing in the opposite direction.

The final piece of his theory was proven with another experiment in which he rotated a bar magnet, with its two ends connected to a galvanometer, around its axis and noticed that an electric current was generated inside the magnet. Obviously, *the magnetic field did not rotate together with the magnet* -- if it had, there would be no changing magnetic field relative to the conductor (the magnet) to produce a current. Once the field was created, it was independent of the source and became a *physical entity permeating space*!

Faraday was forty years old when he made this discovery and he went on for another thirty years experimenting, theorizing, teaching, and producing important discoveries, for example in electrolyzes, but the crowning achievement of his career was done.

Faraday's discovery of the phenomenon of electromagnetic induction and his explanation, using the new concept of electric and magnetic fields, were the final clues needed by Maxwell to unite all the experimental results and theories into his complete and comprehensive Electromagnetic Field Theory, culminating in the famous Maxwell Equations that are constantly used today in science and technology.

James Clerk Maxwell (1831 – 1879)

I can't help noticing that Maxwell was born in the same year that Faraday discovered electromagnetic induction, just as Newton was born in the same year Galileo died. History is full of amusing coincidences.

Maxwell had no doubt about what he wanted to do. Though his father hoped that he would become a lawyer, he chose to study physics and mathematics and was so talented in these areas that his father, after a short hesitation, accepted and supported his decision.

At the age of fourteen, Maxwell produced his first scientific paper on the algebraic treatment of unusual geometric curves, and it was read to the Edinburgh Royal Society. At the age of 16 he was accepted to Edinburgh University where he spent the next three years learning, experimenting, and inventing.

The following four years saw him at Cambridge, doing much of the same, in a much more demanding environment. Here he had competition from the best brains England produced, ncluding Alfred Tennyson, Bertrand Russell, and John Maynard Keynes. Maxwell did very well at Cambridge, both academically and personally. Everybody liked him. He was modest, helpful, compassionate, and deeply religious.

After winning one top and one second prize in a grueling mathematical competition and exams at the end of his four years, his career path was clear and uncontested: Professor of Natural Philosophy first at the University of Aberdeen, then at King's College, London, and, finally, Professor of Experimental Physics at Cambridge University.

At the age of 27. he married the College Principal's daughter, Katherine Dewar. They spent the next 21 years, till the end of his life, in a happy and harmonious marriage. She shared his interest in science and participated in his research in optics. The neighbors across the street watched the two of them suspiciously, as they peered into a coffin-like box in their attic. The box did not contain a corpse; it was their color box, used in optical experiments.

Maxwell had great accomplishments in many areas of physics: color vision and optics (produced the world's first color photograph), astronomy (analyzed the physical composition of Saturn's rings), molecular theory of gases, thermodynamics, and, and most significantly electrodynamics.

In the last, his accomplishments are truly staggering. Ironically, he himself called Ampére the "Newton of Electricity', referring to Ampére's application of the Newtonian "action-at-a-distance" attraction to electrical current segments. However, in the real sense of the expression, Maxwell was the Newton of Electrodynamics, because he performed the same unifying function as Newton had in the area of Mechanics.

Over a span of nine years, he wrote **three separate papers** on the subject, which pulled all of the contradictory and confusing theoretical and experimental results of his predecessors' research into a consistent, integrated theory that unified not only electricity and magnetism but light as well.

1855: *On Faraday's Line of Force*

While most scientists still followed in Newton's footsteps, assuming the unexplained and mysterious "action at a distance" force of gravity, electricity, and magnetism, Maxwell eagerly accepted the alternative explanation offered by Faraday. There is something between the bodies exerting force upon each other, besides the vacuum of empty space.

Faraday imagined invisible lines of force in space, interacting with each other and with material objects in their paths. He could see these lines in the pattern formed by iron filings sprinkled around a magnet, connecting the north and south poles in graceful curves.

This seemed real to Maxwell, much more real than a force being transmitted through nothing, by nothing. However, the lines of force needed precise, mathematical treatment. The first step was making them a continuous distribution over space, rather than Faraday's discrete tentacles of force. There was no point in space through which a line of force would not have passed. This was in agreement with the experimental observation of putting a small test-magnet (or charge) anywhere in the vicinity of a larger magnet (or charge), and always finding a reaction.

This way, he invented the concept of the **field:** a continuous force field that permeated space and became the medium that transmitted the electric and magnetic effects. He believed that this field had a definite structure and tried to describe it by using the analogy of an imaginary weightless and incompressible fluid.

The main contribution to a complete theory of electromagnetism in this paper was Maxwell's concept of the *field* that possessed a definite microscopic structure that could be described for each point in space, by using the mathematical tool of vector analysis. Maxwell read his paper to the Cambridge Philosophical Society in

1855 and received a grateful letter from the aging Faraday, who had almost given up on anyone taking his "lines of force" idea seriously.

However, this paper was just a small step toward explaining the whole range of observations about electrical and magnetic phenomena accumulated before.

1861: *On Physical Lines of Force*

The next step had to wait six years, during which Maxwell was very busy with his work on Saturn's rings, a kinetic theory of gases, color vision research, and marriage. Finally, in 1861 he published the second paper on his electromagnetic theory wherein he predicted the existence of electromagnetic waves. It was an incredible leap forward, foreshadowing the entire communications industry of today.

In this paper, Maxwell wanted to move beyond the arbitrary analogy he had used before and attempted to provide a hypothetical mechanical model for the structure of the electromagnetic field. Without a shred of experimental data, Maxwell constructed an elaborate mechanical clockwork model with invisible spinning cells and rotating, moving wheels that permeated all space. Yet, when he developed his mathematical equations, they described the magnetic Coulomb Law, the magnetic induction effect of an electric current and the induction of an electric current by a changing magnetic flux.

Maxwell published his paper in the *Philosophical Magazine*, spread over the March, April, and May 1861 issues. He made sure it was understood that his model was arbitrary, a result of his imagination and not necessarily describing nature.

However, Maxwell was disappointed with his theory not being able to explain the electrical Coulomb Law and kept thinking about this last challenge, until he had an idea that might work. If he allowed his cells to be elastic, their elasticity could explain the force between electrically charged objects. When he developed this idea mathematically, it all worked out.

The most important consequence of this theory was the prediction of **electromagnetic waves**. Since all elastic materials can transmit waves, Maxwell's elastic cells were able to transmit a disturbance at one point in such a way that "any change in the electric field would be accompanied by a change in the magnetic field and vice versa". The waves would be transverse: the disturbance is perpendicular to the direction of the wave.

According to Maxwell's calculations, the speed of this wave in vacuum was the same as the ratio of electromagnetic and electrostatic units of electrical charge. This ratio had been measured by Weber and Kohlrausch in 1856 to equal the known speed of light (see later).

Maxwell refused to believe that this was a mere coincidence. The irresistible conclusion for him was: light itself consists of electromagnetic waves.

Maxwell published an extension to his paper, *On Physical Lines of Force,* in two more parts in 1862. The essence of the now complete theory was the first mathematically treated mechanical model of the aether - the medium through which both electromagnetic waves and light could travel, and the two were one and the same.

Reactions to the paper were mixed. The ideas were so startling that many scientists expressed misgivings. His friend, Cecil Munro wrote to Maxwell: "The coincidence between the observed velocity of light and your calculated velocity of a transverse vibration in your medium seems a brilliant result. But I must say I think a few such results are needed before you can get people to think that every time an electric current is produced a little file of particles is squeezed along between two rows of wheels."

Cecil Munro's wish was granted: in 1887 Heinrich Hertz experimentally demonstrated the existence of electromagnetic waves.

1864: *Dynamic Theory of the Electromagnetic Field*

Most of Maxwell's contemporaries expected him to develop his clockwork model further and describe and predict other startling properties of the light and the electromagnetic wave-carrying-aether.

They were disappointed. Maxwell abandoned his model and struck out in a completely new direction that did not contain any mechanical analogy but, instead, used only known principles of dynamics of matter and motion.

Without a mechanical model, Maxwell had to treat the electromagnetic system as a "black box", where only the input and output were known, without any idea of what happened inside the 'box'. By that time Maxwell realized that we may never understand the detailed workings of nature and may have to be content with describing the relationship between what went in and what came out.

The result was a set of 20 equations on 20 variables. After several revisions by Maxwell and others, these were reduced to four vector-equations that describe everything we know about electrodynamics and they provided many predictions and explanations that were unknown at the time.

Maxwell's equations describe the **structure** of the **electromagnetic field** at each point in space, in terms of what mathematicians call "**divergence**" and "**curl**".

If a force field has a positive **divergence** at a point in space (often called a 'source' or 'faucet'), it means that the forces point outward from that location in space, like the repulsive force of a positive charge against all other positive charges. If it has a negative divergence (often called a 'sink' or 'drain'), then the forces point inward to that location, like the gravitational attractive force of masses.

On the other hand, if a force field has a **curl** (sometimes called a 'whirlpool') at a point in space, then the forces curl around that location, like the magnetic forces around a current-carrying wire.

Maxwell had two equations for the electric field (divergence and curl) and two equations for the magnetic field (divergence and curl). The equations incorporate the basic experimental discoveries in a precise mathematical form.

The first equation is deduced from Coulomb's law and the superposition principle by using a mathematical vector analytical formula called Gauss's law. It states that the divergence of the electric field surrounding a point equals the electric charge density at that location because the source of the electric field is always electric charge. No new physics is expressed in this law: it describes the consequences of the Coulomb Law for any point in space and time for the electric field, as defined by the charge density at the same location and moment.

The second equation states that the curl of an electric field around a point equals the rate of change of the magnetic field at that location. Faraday's induction experiment was reformulated as the structure of a vector field: the induced electric field at one point in space and time depends on the rate of change of the magnetic field at that location and moment.

The third equation has no special name. It states that the divergence (source) of a magnetic field at any point is zero, due to no magnetic monopole ever being observed.

The fourth equation states that the curl of a magnetic field around a point equals the sum of the current density and the rate of change of the electric field at that location. The first part reformulates Ampére's force law for each point in space and time for the induced magnetic field. The second part: the rate-of-change-of-the-electric-field term was Maxwell's suggestion, based on the assumption of symmetry: since the change of electric field resulted in an induced magnetic field (second equation), he thought that the reverse should be true as well: any change in the magnetic field ought to induce an electric field.   Maxwell was absolutely right in this supposition, and it was experimentally verified later.

These four equations, together with the 'Lorentz Force Law' and the assumption of the conservation of electric charge covers everything then known about electrodynamics.

If we know the charge density and current density in space, as a function of space and time, then, with the Maxwell equations, we can find the value of the electric and magnetic fields, also as a function of space and time. His contemporaries were baffled, so new and so revolutionary was this idea. Time would soon prove Maxwell right.

Lorentz Force Law

Experiments have shown that if we suspend a conducting wire between the poles of a U-shaped magnet and start an electric current through that wire, a force will be exerted on the wire which is perpendicular to both the direction of the current and the line connecting the poles.

In the *Treatise on Electricity and Magnetism*, published in 1873, Maxwell describes this experimental result and gives a formula that describes this force. This formula later became known as the **"Lorentz Force Law",** due to its generalization by Dutch physicist Hendrik Antoon Lorentz.

We know that the current can be looked at as electric charges continuously flowing through the wire, so we can assume that the magnet exerts a force on moving charges. Measuring this force, we find that it is proportional to the strength of the magnet, the quantity of the electric charge, the speed with which it is moving and the angle between the wire, and the direction of the magnetic lines of force.

$$F=k_L Qv \sin \zeta$$

where 'F' is the measured force, 'Q' is the electric charge moving with speed 'v' and 'ζ' is the angle between the direction of the wire and the magnetic field.

This Lorentz Force appears any time an electric charge moves with a definite velocity relative to a magnetic field, regardless of how the field is generated. The field can be due to a bar magnet or an electric current through a conductor, the magnetic field of Earth, or of unknown origin.

We use the Lorentz force law to define the magnetic field: we say that there is a **magnetic field** present if we observe a force acting on a moving electric charge and that force is not due to the electrostatic force exerted by an electric field. In this case, both the direction and the magnitude of the magnetic force is given by the Lorentz Force Law.

Measurements

Most of the knowledge gained before Maxwell was qualitative rather than quantitative. Different scientists used different ways, and chose different units, to measure electric charge, magnetic momentum, electric current, and resistance. It is amusing for us today that great scientists used frog-legs for measuring electricity when Volta invented his battery.

Before that, other scientists used the tingling sensation on their tongues to see whether a weak electric charge was present, which was not as extreme a method as the one used by Cavendish who gave himself electric shocks to determine, from the measure of pain, the relative charge of a Leyden jar. Due to this inaccuracy, it was very difficult to notice underlying connections.

Maxwell was keenly aware of the need for precision and quantitative (mathematical) treatment of the emerging science of electrodynamics and helped put it on a solid practical foundation. He wrote a paper on the subject in 1863 recommending a complete system of units for electricity and magnetism. His recommendations were accepted and became known as the Gaussian System.

According to an article written by British physicist Richard Clarke of the University of Surrey: "Units are no longer taught extensively. Their bland and pedantic nature makes the study of drying paint more exciting." It is a subject that makes students' eyes glaze over, certainly.

However, without selecting unit values for the new quantities in the electromagnetic formulas, the science remains a theoretical discussion without any practical value.

Remember the electromagnetic formulas we discussed earlier:

	Electricity Magnetism	Induced Magnetism	Force on moving charge
Coulomb	$F=k_e \, q_1 q_2/r^2$ $F=k_m \, p_1 p_2/r^2$		
Biot-Savart		$df = k_B \, I \, ds \, \sin \zeta \, / \, r^2$	
Ampére		$F = k_A \, I_1 \, I_2 \, L/r$	
Lorentz			$F=k_L Q v \sin \zeta$

Every formula contains a proportional factor, the value of which can be determined only by designating a unit value for the newly discovered concepts such as electric charge, magnetic pole strength, and electric current. Once this is done, precise measurements can determine the value of the k_e, k_m, k_B, k_A, and k_L constants. It will be explained in detail in the "Next Level" chapter.

Precise and consistent use of unit values can unmask hidden connections of tremendous significance.

Who has not heard of Einstein's famous formula $E = m\,c^2$, where 'c' is the speed of light in a vacuum? How did the 'c' get into the formula? Some students would know that this 'c' is from Maxwell's equations, but very, very few know how the speed of light got *into* those equations in the first place.

The answer is in the experiment performed by Weber and Kohlrausch in 1856, to demonstrate the relationship between measured electrical constants k_e, k_m, and k_B. To their surprise, they found that the square root of the measured value for the $k_e k_m / k_B^2$ constant equals the measured value for the speed of light. At the time, most scientists thought it coincidental, but this experimental result reveals a hidden connection between electricity, magnetism, and light. It provides a significant clue to both Maxwell's famous equations and Einstein's Special Theory of Relativity.

Apart from obscure connections discovered through precision in measurement, it is very important, especially for students, to understand the origin and of the terms used in electrodynamic formulas, and why they are used. For example, in modern books on Electrodynamics, Biot-Savart's law contains the following term: $1/4\pi\varepsilon_0 c^2$. Why 4π? What is 'ε_0'? Why 'c^2'? Either the student learns by rote, or he knows and understands these symbols and what they represent. We will deal with these symbols in the "Next Level" chapter when discussing measurement systems in detail.

Heinrich Hertz (1857-1894)

The word 'hertz' is known worldwide as the unit of frequency of electromagnetic (or radio) waves. Yet, the tragically short life of Heinrich Rudolf Hertz, a German physicist, is relatively unknown.

He is not famous for discovering any new basic principles like Oersted and Faraday, nor for synthesizing decades- or centuries-long research into a complete, consistent theory like Newton, Maxwell, and Einstein. What Hertz did was perform the vital experiment that proved Maxwell's prediction of electromagnetic waves and opened up the area of technology on which we utterly depend today. This includes communications, the entertainment industry, and any subsequent invention that uses electromagnetic waves for wireless transmission.

Hertz used two loops of wire. One - the transmitter loop – passed through a switch and had its two ends connected to a battery. It was placed at one end of the room. The other - the detector loop – had a small gap between its ends and no source of power. It was placed at the far end of the room. When the switch was turned on and current flowed through the transmitter loop, a small spark jumped across the gap in the detector loop. As the detector loop was moved about in the room, the intensity of the sparks varied in proportion to its distance from the transmitter loop. The sparks were caused by electromagnetic waves transmitted from the first loop to the second, without contact, and thus Maxwell's prediction was fully verified.

Within a short time, Hertz's results were repeated by other physicists around Europe and it did not take long for practical applications to come along. In 1894, Italian electrical engineer Guglielmo Marconi (1874-1937) came across an article on the "Hertzian waves", as they were then called, and with increasingly refined experimental setups he managed to communicate Morse code in a wireless message over greater and greater distances. On December 12, 1901, he successfully transmitted radio signals from Europe to North America (Poldhu, Cornwall to St. John's, Newfoundland). Radio, radar, TV, radio astronomy, GPS, the Internet, and all other technologies using electromagnetic radiation were born.

The discovery of the electron (1897)

Prior to the discovery of the electron, there was hardly any agreement regarding the nature of electricity and magnetism. In Maxwell's time, atomic theory was only a contested hypothesis. Nobody as yet had a clue what electric charges were.

Since it could demonstrably 'flow' from one place to another, many considered electricity to be some kind of 'fluid' inside matter. Arguments concerning this theory disputed the nature and number of these fluids. Electric and magnetic attraction and repulsion were thought to be similar to Newton's force of gravity, acting instantaneously over a distance, without any intervening material contact. Few took Faraday's idea of invisible forces filling out space seriously. Maxwell was one such.

Basic assumptions made by key players were as follows:

	Electricity	Magnetism
Coulomb	Two fluids (+,-)	Two fluids (astral, boreal)
Benjamin Franklin	One fluid (+) (access and deficiency)	One fluid (+) (excess and deficiency)
Biot-Savart	Two fluids (+,-)	Two fluids (astral, boreal)
Ampére	Two fluids (+,-)	Two electric fluids (+,-) electric in nature
Oersted	Forces	forces
Faraday	Electric lines of force	Magnetic lines of force

The Maxwell equations were the crowning achievement of centuries-long research into electromagnetic phenomena, in the sense that it gave a precise, mathematically consistent description of the structure of the electromagnetic field, and provided the tools needed to calculate the value and behavior of the electric and magnetic fields, depending on the charge and current distribution.

However, it said nothing about the *nature* of electricity and magnetism; provided no mechanism to account for electric currents or magnetic effects. Such a mechanism became available only after J.J. Thomson's discovery of the electron in 1897.

Science never follows a straight path from discovery to theory.

The significance of Electrodynamics

Now that we have covered the basic concepts, definitions, and laws of electrodynamics, we need to ask ourselves: what does it mean in terms of our view of the world? The practical consequences are obvious and impressive: countless conveniences in the life of our species and our civilization.

What has electrodynamics given us in the philosophical sense? Is it comparable to the spiritual/philosophical revolution started by Copernicus and completed by Kepler, Galileo, and Newton? Nothing so dramatic as Galileo's trial happened as a consequence of electrodynamics. The effect it had on our thinking was more subtle. It reinforced our 'faith' in science -- a 'faith' first established by Newton's great success in explaining the cosmos. After the triumph of electrodynamics, we expected no limits to the power of the human mind.

Ironically, while we basked in the glory of our accomplishments, the seeds of doubt were already being planted by the next genius: Albert Einstein. However, that was still in the future.

At the end of the 19th century, we had electricity in experimental labs, in theoretical textbooks, in physicists' minds. One more major step was required before it became the property of all mankind: industrial application. For that step, we needed large-scale electricity generation and distribution. It may sound like a straightforward engineering task, but in fact many ingenious inventions were required before ushering in the electrical age.

For that final step, one more fundamental science had to mature: Thermodynamics. Without the basic principles of thermodynamics, we could not have had a practical and efficient source of electricity. All of our early power-generating capacity was based on some form of heat engine that required steam to power rotating turbines. Thermodynamics teaches us how to build maximum-efficiency heat engines.

Chapter VI - Thermodynamics

Cause-and-effect chains in human history tend to be very long. They often start with a search for, and acquisition of, new knowledge about our physical environment. New knowledge is the result of the labor of thousands of inquisitive minds collecting data over decades or centuries until it reaches critical mass, and then a very few of humanity's creative geniuses condense all this accumulated information into coherent theories.

The rest is up to the engineers and technicians who take possession of new science and turn it into technology, radically altering our environment. We create cities with larger and larger buildings, schools, hospitals, transportation, and communication networks.

Due to the benefits of advancing technology and the human population grows exponentially, swarming over every liveable square mile of the planet, affecting all of it. We contribute to climate change, pollution, the depletion of resources, and the extinction of countless species of animals, insects, fishes, and plants.

We also create new species with genetic engineering and, by mapping the human genetic code, we hope to eliminate most human diseases and maybe even dramatically increase our longevity. We venture into space, think about colonizing other planets, search for extraterrestrial intelligence and study the mysteries of the expanding universe with optical and radio telescopes stationed on Earth, in orbit, on other planets, and deep space probes.

While rushing forward on all these fronts, we spend our days in ever-growing fear of the super bombs, the superstorms, and the superbugs we may have unleashed with our uncontrolled charge into an unknown future.

The waves of advancement in knowledge and understanding often start in antiquity and result in our current technical civilization. These waves overlap to a large degree during their initial phases, run parallel for a while, influence and inspire each other and reach their peaks in no particular order, usually when all the relevant data has been collected and the topic is ready for a consistent, coherent theory to explain all (or at least most) of the accumulated observations.

The first well-documented wave of this process started in ancient Greece at about 600BC and, though interrupted by a 1000-year Dark Age, culminated in Newton's synthesis in 1687. The primary focus of this wave was understanding the laws of motion both here on Earth and in the heavens. The result of this wave was the science of Mechanics. It enabled engineers to build clever machines that affect all aspects of civilization. On the spiritual front, it led to the decline of religion and the ascent of science, at least in the western mind.

The second wave started in the 18th Century with the study of electricity and magnetism, and resulted in the science of Electrodynamics, based on Maxwell's equations. This science gave us the modern communication industry and revolutionized all other aspects of civilization from medical science to education and entertainment. Spiritually this wave brought us a hint of a universe that may be incomprehensible to minds that evolved in only a very limited segment of reality. Einstein's Relativity Theories shook our cocky complacency for the first time in the three hundred years since Newton's triumph in explaining the then visible universe.

The third wave in the quest for knowledge (my apologies to Alvin Toffler) is as old as mankind itself and is still rolling over the landscape of science and engineering today. It concerns our most fundamental, most immediate needs for survival: matter and energy.

We need to understand the material world to find out what matter is made of, how it can be manipulated and transformed to create the artificial environments on which we depend. We need to find sources of abundant and clean energy to heat our houses, cook our meals, run our vehicles, power our factories and farms.

The quest for understanding matter started with the labor of chemists who studied the properties of elements and compounds and the laws of chemical interactions. It resulted in more and more detailed knowledge of the building blocks of matter: atoms, electrons, protons, neutrons and, quarks. This line of inquiry created chemistry, followed by atomic and nuclear physics.

Our quest to secure clean energy for our survival entered a critical phase. We are fast running out of fossil fuels, waging brutal imperial wars for the last drops of oil, and drowning ourselves and our world in chemical pollution - jeopardizing our very survival through reckless, unsustainable expansion in the wrong direction!

We have only one hope left; the ultimate prize science can offer mankind: clean, abundant, inexpensive energy. The first step along this route, nuclear fission, was a false start: it is not clean, and it is dangerous. One practical alternative is nuclear fusion. Given the will, we can make it work. We know how to liberate the energy locked inside matter (in our H-bombs) we 'only' need to do it in a safe and sustainable way. We have been living off the power generated by nuclear fusion through all our history: the energy coming from the sun, which is the source of all life on Earth. We need to create our own reliable and safe 'sun' to provide more of what we need.

On the spiritual front of this third wave, the almost totally incomprehensible world of the very small resulted in a philosophical crisis. The "uncertainty principle" discovered by Heisenberg gave us the first hint of a possibly unsurpassable limit to human understanding.

The subject of this Chapter is the beginning of the third wave.

What can it do for me?

Engineers usually take over from physicists after the crucial experiments have been performed and some kind of rudimentary theory has been worked out. It certainly happened like that in electrodynamics.

In thermodynamics, the process was, uncharacteristically, reversed. Thermodynamics, as a science, was started by an engineer who wanted nothing more than to develop the perfect steam engine.

His name was Nicolas Léonard Sadi Carnot (1796 -1832), a French military engineer of Napoleon's reign. In 1824 he published his only work *On the Motive Power of Fire,* in which he investigated the maximum work/heat ratio that could be obtained from a heat engine. He was the first to describe the relationship between heat and work in a quantitative way. His theory inspired Clausius, 25 years later, to formulate the "Entropy Law" which is the essence of what we call today the "Second Law of Thermodynamics".

Progress, in any discipline of science, inevitably starts from a state of confusion. When studying a phenomenon such as heat and heat-related processes, we already know a large number of things and had been theorizing, often wildly, about their hidden causes. In thermodynamics, it did not take us long to learn about temperature, heat, work created by heat engines, and how heat seemed able to travel inside bodies and even be created by mechanical means (friction).

The confusion at the beginning consisted of not knowing how these known quantities related to one another. What was heat? Was it like a fluid that could flow? Was it indestructible like mass? Could it be created from other sources? Was there an explanation for heat from the debated atomic structure of matter?

With the invention of the steam engine, humanity started a long line of devices that convert heat to mechanical energy, which can then be used to do physical work. The common name for these devices is "heat engines".

For practical reasons, the study of heat started with the study of 'heat engines. Partly, it was important to perfect them for the economy; partly because one particular type of heat engine, the steam engine, was easy to analyze. The mysterious process of heat causing mechanical work happened inside a chamber of the engine that was filled with a gas: high temperature, high-pressure steam which caused the piston to move and produce work.

Other types of heat engines familiar to us today are, for example, the internal combustion engine in cars and the heat exchanger in geothermal heating systems in homes.

One thing common to all heat engines is that they use heat from burning some fuel (coal, oil, gasoline, uranium) or the stored heat in the Earth or oceans, to raise the temperature of a suitable substance (usually gas or liquid, such as water) in a container, to a high temperature, and then use this substance to produce mechanical work while losing some of its temperature.

Gases are easy to study - all you need is a cylindrical container filled with the gas, a heat source outside, and the mechanics attached to the piston inside the cylinder, so the volume and pressure of the gas inside can be changed.

Most of the early discoveries were made by experimenting with gases since it requires minimal equipment. Some of the more frequently used gases:

Fluorite: In 1529, Georgius Agricola described fluorite as an additive used to lower the melting point of metals during smelting.

Hydrogen: In 1671, Robert Boyle discovered and described the reaction between iron filings and dilute acids, which results in the production of hydrogen gas.

Nitrogen was discovered by Scottish physician Daniel Rutherford in 1772, who called it noxious air.

Oxygen was first discovered by Swedish pharmacist Carl Wilhelm Scheele. He had produced oxygen gas by heating mercuric oxide and various nitrates by about 1772.

Chlorine was first prepared and studied in 1774 by Swedish chemist Carl Wilhelm Scheele.

Before getting further into heat engines, we have to inventory the huge volume of accumulated data regarding energy, temperature, heat, and heat processes before the theory could be advanced.

First Steps

Human beings are the most self-centered creatures in the universe (as far as we know). We measure and judge everything we encounter according to how it affects us, how it compares with our life experience. That is why relativity theory is so hard to understand and accept: it doesn't compare very well with our limited experience of speed.

Remember how Cavendish gave himself electric shocks to determine, from the measure of pain, the relative charge of a Leyden jar? Newton almost went blind by poking at his eyeball with a blunt needle to see how his vision was affected by the distortion to its curvature.

Naturally, the concept of temperature was born out of our physical sensation of 'warm' and 'cold'. That was a very crude and unreliable way to determine temperature. Everyone is familiar with the experiment of putting one hand into hot, the other into cold water, and then moving both hands into a basin of lukewarm water: one hand tells us it is cold, the other tells us it is warm. An instrument was needed to measure temperature accurately.

It was recognized that objects change their dimensions in response to temperature: they expand with heating and contract with cooling. It was also recognized that when two bodies of different temperatures are brought into contact with each other, in time, their temperatures equalize.

These two observations led to the invention of thermometers. Since expansion and contraction happen in a reliable and objective way, scientists decided to measure temperature by observing how much expansion or contraction occurred when a suitably chosen testing object was brought into contact with a test object of unknown temperature.

In 1592 Galileo invented the first crude thermometer: a glass tube, with a closed glass bulb at one end, the other end open. He half filled it with water and then, holding it vertically over a shallow pan of water, placed the open end into the pan. Since the bulb-end was closed, the water stayed in the tube due to atmospheric air pressure on the water in the pan. As he heated the bulb, the gas expanded and pushed the water level down in the tube. When he cooled the bulb, the opposite happened. With this device, he could indicate the temperature of an object brought into contact with the bulb.

In 1662 British physicist Robert Boyle discovered an inverse relationship between the volume and the pressure of a gas: $pV = const$. French physicist Edmé Mariotte discovered the same law independently 16 years later, adding the important observation that it is true only if the temperature is kept constant. It is called the Boyle-Mariotte Law today.

In 1699 French physicist Guillaume Amontons invented the thermometer that used the change in air pressure, (rather than volume, as Galileo did) to measure the change in temperature. With this device, he found that water always boiled at the same temperature and the volume of gas increased with temperature at the same rate, regardless of the kind of gas (oxygen, nitrogen, etc.) he tried it on.

In 1714 German physicist Gabriel Fahrenheit invented the closed thermometer (isolated from outside pressure) with a mercury column that expanded and contracted in a precise and reliable way with the change in temperature. He knew that the lowest temperature he could reliably attain was the mixture of ice, water, and ammonium chloride. To be able to define a scale he needed another known and reproducible temperature and he chose the melting point of ice. He decided to call the lowest temperature 0 degrees and the higher 32 degrees. These two values defined the scale by dividing this difference into 32 equal parts. According to this scale, the temperature of boiling water became 212 degrees in the Fahrenheit scale.

In 1742 Swedish astronomer Anders Celsius suggested another scale in which the zero point was determined by the freezing point of water and the boiling point of water was chosen to be represented by 100 degrees. This is commonly used today as the Celsius Scale all over the world, except in the USA.

With a reliable, repeatable, and objective way to measure temperature, the stage was now set for studying other temperature-related phenomena.

As Albert Einstein wrote (with Leopold Infeld) in *The Evolution of Physics*:

"A pound of water placed over a gas of flame takes time to change from room temperature to the boiling point. A much longer time is required for heating twelve pounds, say, of water in the same vessel, by means of the same flame. We **interpret** this fact as indicating that now more of "something" is needed and we call this "something" – *heat*."

Please observe carefully how science works. We decide what properties we want to measure with the experiment (for example volume, pressure, and temperature in a container of gas). What we measure is clearly defined by the measuring process. We measure these properties before we start the experiment. Then we perform the experiment and measure the same properties again, in the exact same way. Most often we find that the experiment changed one or more of the measured properties. Then we make assumptions about things we cannot see but assume to cause the observed result.

In electricity, for example, we measured the force of attraction or repulsion between two electrically charged objects at different distances from each other. The methods for measuring the effect of force are standard practice in physics (Coulomb used his own invention: a torsion balance) and everyone can see what we are measuring. Then we make an assumption about an invisible quantity we call "electric charge" that must be the cause of this attraction. We also assume that this "charge" can flow from one object to another and when two identical objects are touched together, we assume that the charge will be identical on both and half the value of the original. This way Coulomb was able to come up with his formula for electrostatic attraction. He still did not know what "charge" actually was, but now he could measure it by using an arbitrarily created test charge he called the unit charge.

In thermodynamics, the process was very similar. We could measure the heat expansion and contraction of a mercury column and considered this phenomenon an indication of its "temperature". Then we assumed that there is something responsible for this temperature and we called it "heat". This "heat" can obviously flow from one object to another, since in our experience temperatures even out. So now we can study this hypothetical phenomenon, heat, and determine its properties. Even though we have never seen it or touched it, we have described what it does.

It is a bit like that old joke:

Query: What was there before God said "Let there be light"?
Answer: Nothing
Query: What was there after?
Answer: "Still nothing but now you could see it"

In 1760, Scottish chemist Joseph Black performed some important experiments. He heated equal weights of mercury and water over the same flame and found that the temperature of mercury went up twice as fast as that of water. From this experiment, Black assumed that twice as much "heat" was needed to raise the temperature of water by the same amount as the temperature of mercury. He introduced a new concept: "**heat capacity**" or "**specific heat**". This is the quantity of heat that must be added to 1 gram of the substance to raise its temperature by 1 degree Celsius. Knowing the heat capacity allowed him to invent a rudimentary device to measure the heat content of any object.

The device was called an **ice calorimeter.** It consisted of a container placed inside a larger insulated container. The gap between the two containers was filled with ice. The test object was placed inside the inner container and, after a suitable time, the mass of the ice melted between the two containers was measured. This gave an indication of the heat content of the measured object.

The calorimeter enabled him to define the unit of heat, the "**calorie**", thus: One calorie of heat is the amount of heat required, at a pressure of 1 Atmosphere to raise the temperature of 1 gram of water by 1 degree Celsius (specifically from 15 to 16 degrees).

Different materials with different specific heat require different amounts of heat to raise their temperature, according to the following experimental formula:

$$Q = c_v\, m(T_2 - T_1)$$

where c_v is the specific heat of the material studied, 'm' is its mass, 'T_1' is the starting temperature and 'T_2' is the final temperature. Q is the heat, measured in calories or "cal". According to this formula, the precise definition of specific heat of a substance is:

$$c_v = Q/m(T_2 - T_1)$$

which is the heat required to raise the temperature of 1 gram of the substance by one degree Celsius.

You will have noticed one difference between the concept of "heat" in thermodynamics and the concept of "charge" in electrodynamics. The measurement of electric charge gives us the total charge of an object, while we can only measure increments (positive or negative) of heat, without any notion of what the total "heat-content" of an object is. This is because we can easily find objects with zero electric charge (no force measured by a testing device) but we don't know how to find an object with zero heat content.

In 1762, Black made further progress. Heating a mixture of ice and water, he noticed that the temperature of the mixture did not change until all the ice had melted. The same thing happened when water was boiled: the temperature did not change until all the water turned to steam. Black called the form of heat that did not result in an increase of temperature "**latent heat**". Black found that different liquids required different amounts of latent heat.

All of Black's experiments suggested that heat had to be some kind of substance that came to be called "caloric". This is similar to the equally erroneous assumption of a "magnetic fluid" that we encountered in electromagnetic research.

Ideal Gas Law

The earliest steps in thermodynamics culminated in one of the most useful concept: the "Ideal Gas" and its properties.

In 1802, French chemist Gay-Lussac discovered a relationship between the volume of a gas and its temperature. If pressure is kept constant, the volume increases or decreases with temperature, according to the following formula:

$$V = V_0(1 + \beta t)$$

where 't' is the temperature of the gas and 'β' is a constant value, the same for all gases: $\beta = 1/273$ and V_0 is the volume of the gas at zero degrees Celsius. This implies that, in theory, the volume of a gas would shrink to zero at the temperature of -273 Celsius – in reality, it will turn to liquid and solid long before that.

The Boyle-Mariotte Law and the Gay-Lussac Law gives us the **Combined Gas Law.** To make calculations easier, physicists invented a hypothetical gas, called "ideal gas", for which the combined gas law is valid for any temperature and pressure.

Let's assume that at zero Celsius temperature the pressure of a gas is p_0 and its volume is V_0.

If we keep the pressure constant p_0 and change the temperature to 't' then, according to the Gay-Lussac Law the volume will change to:

$$V' = V_0(1 + \beta t)$$

If now we keep the temperature constant 't' and change the volume to 'V' then, according to the Boyle-Mariotte Law

$$pV = p_0V'.$$

Substituting the value, we had for V' into this will give us:

$$pV = p_0V_0(1+\beta t)$$

where 'p' and 'V' are the pressure and volume of the gas at temperature 't' Celsius degrees and 'p_0' and 'V_0' are the pressure and temperature of the gas at zero Celsius degrees and the value of $\beta = 1/273$. Substituting the value $\beta = 1/273$ we can rewrite the equation as:

$$pV = p_0V_0(273 + t)\,/273$$

and introducing the $T = t+273$ variable and the $T_0 = 273$ constant (temperature of freezing water in Kelvin degrees) we get:

$$pV/T = p_0V_0\,/\,T_0$$

or

$$pV = CT$$

where $C = p_0V_0\,/\,T_0$ value, for a given amount of gas, depends only on the kind of gas we are using (at constant temperature T_0 the p_0V_0 value is a given for each gas, according to the Boyle-Mariotte law).

Obviously, the $p_0V_0\,/\,T_0$ value is proportional to the amount (mass) of gas we are using, because if we connect two containers of (p_0, V_0, T_0) gas, then p_0, and T_0 won't change, but both the mass and the volume will be doubled.

Therefore, the $C = p_0V_0\,/\,T_0$ constant can be written as:

$$C = Rm$$

Where 'm' is the mass of the gas and 'R' is independent of the amount of gas and depends only on the kind of gas we are using. This gives us the combined gas law as:

$$pV = mRT$$

This gas law is valid only within a limited range. The smaller the pressure and higher the temperature the more accurate it is.

Count Rumford (Benjamin Thomson) 1753 - 1814

Earlier I wrote: "As Copernicus has been called the most colorless of the great scientists of the Renaissance, Tycho de Brahe was probably the most colorful". Well, Count Rumford is at least as colorful as Tycho was, even though the two had little in common.

G.I. Brown titled his biography of Rumford *Scientist, Soldier, Statesman, Spy,* and every one of the claims is absolutely true. Historian Edward Gibbon, who was fascinated by this versatile American, referred to him as "Mister Secretary, Colonel, Admiral, Philosopher Thompson".

Count Rumford was born in 1753 with the much more modest name of Benjamin Thomson to humble farmers in Woburn, Massachusetts. Mostly self-educated, he began his career by tutoring the children of wealthy families, then teaching in a small school in Concord, New Hampshire. He married his patron's daughter, the widow of a rich older man. She was thirty to Benjamin's nineteen years of age. Through Sarah's connections, he was able to obtain the rank of major in the 2nd New Hampshire Regiment of the British army.

The American War of Independence forced him to choose sides and he opted for Britain. He served by spying on the rebels and passing military information on to the British. When discovered, he had to flee to England, leaving his wife and young daughter behind.

With glowing references from the British military commander in Boston, he found ready acceptance. However, when the peace treaty was signed by Britain and America in 1782, Benjamin found that his services were no longer required.

As a career soldier, he looked around for trouble-spots in Europe, eventually making his way to Germany. He was almost immediately championed by the Elector of Bavaria, Carl Theodore, who promoted him to colonel in the Bavarian army. After a brief visit to England, where he was knighted for his prior services, he went back to Munich. After a few years of distinguished service, he was appointed Minister of War and Minister of Police in 1788.

He immediately set out to reorganize the Bavarian army. He solved the town's terrible problem of beggars and thieves by arresting them and employing them gainfully in workhouses, manufacturing uniforms for the army. He did all this with such ingenuity that everybody, most of all the underclass, benefited handsomely.

It is unclear whether his motivation was compassion for the poor or, as I suspect, the mindset of an engineer who could not stand waste and wanted to produce maximum results with minimum cost in resources and assets. He reminds me of Hank Morgan in Mark Twain's *A Connecticut Yankee in King Arthur's Court*. When Hank Morgan sees a hermit standing on top of a cliff, bowing deeply for hours on end in prayer, he wonders how all that wasted energy could be used. He then comes up with the bright idea of attaching a harness to the hermit's body to power a sewing machine, making 'hermit-t-shirts' to sell to the pilgrims.

Thomson didn't just give orders; he actually got his hands dirty, experimenting with everything from the heat-conducting qualities of different textiles to the nutritional requirements of large masses of people. His projects were a major success and, as a reward, he was knighted again and became "Count Rumford" of Bavaria.

One more feather in his cap was as commander of the Bavarian army in 1796. By playing off the French and Austrian armies against each other, he saved Munich from invasion and became a national hero. As the war on the continent ended, Rumford found little to satisfy his ambitions, so decided to return to England in 1798.

This very busy man, during all these adventures, never stopped thinking, observing, experimenting, and writing about his ideas on a large variety of subjects. One research paper he submitted to the British Royal Society in 1792, on the nature of heat, which was not well understood at the time, received their highest honors, the Copley medal.

His most significant contribution to science was demolishing the theory of heat-fluids. While boring cannons for the Bavarian army, Rumford noticed that the metal grew so hot as the boring tool gouged it out, so they had to be cooled constantly with water.

Evidently, heat had something to do with motion. Not only that, but it seemed that enough heat could be generated by friction that, had that heat been present in the brass, it would have melted the metal before the boring ever began. Therefore, the heat had to be produced by the friction; had to be the result of mechanical motion.

The debate over the exact nature of heat was in full swing when Rumford made his crucial experiments and announced his results in a second paper submitted to the Royal Society in 1798, titled *An Experimental Inquiry Concerning the Source of Heat Which is Excited by Friction*.

One of the recurring themes in science is the quest for understanding what things are made of and how they work. It has come up over and over in the history of science: the planetary system, gravity, light, sound, electricity, magnetism, matter. We have senses to observe all of these things but often it is not obvious what we are observing.

So, it was with heat and Rumford made a major contribution.

Sadi Carnot (1796-1832)

In 1824 French military engineer Sadi Carnot examined the theoretical limits to the efficiency of heat engines.

First, he compared the heat engine to a water mill, where the fall of water from a high potential energy level to a lower one produces work in the process. Using this analogy, Carnot found the common elements in any kind, not only steam, of heat engine: heat passes from a high-temperature level (the boiler) to a lower temperature (the cooler), producing useful work in between. No heat was lost in the process, he thought, the same way as no water is lost in the watermill. This consideration suggested that the limit to the efficiency of any heat engine is determined by the two temperatures.

Next, Carnot tried to estimate the limit by imagining a perfect engine, without internal friction, poor insulation, or any other cause of heat loss. In such an engine, the process of transporting heat from a hotter to a colder reservoir, and doing work in the process, can be reversed: we could put in work and make the heat flow 'up' from the colder to the hotter reservoir. (as, in effect, a refrigerator does.) Carnot called his imaginary engine **"reversible"**.

Carnot's third great contribution was what is now called **"Carnot's Theorem"**: the efficiency of a reversible engine depends only on the temperatures of the two heat reservoirs. This means that if the boiler has an absolute temperature of T_1 Kelvin degrees and the cooler has an absolute temperature of T_2 Kelvin degrees, then the rate of mechanical work 'L' and the amount of heat 'Q' input from the boiler is determined by the following formula:

$$L/Q = 1 - T_2/ T_1$$

According to this formula, the maximum efficiency (perfect conversion of heat to work without any loss) can happen only if $T_2 = 0$, that is the cooler would be at absolute temperature zero (-273 Celsius).

His insight proved correct. In possession of the first law, it is easy to demonstrate by analyzing the four phases of the "Carnot-cycle" of a reversible engine, using ideal gas as the medium to deliver heat from the boiler to the cooler and produce mechanical work in the process.

On old-fashioned steam locomotives, the arm of the piston, moving in and out of the steam-filled cylinder, is highly visible. We see how this arm is driving the wheels of the train - that is, doing mechanical work. The operation of this engine is cyclical; it repeats the same action over and over.

For the casual observer, there are only 2 phases in the cycle: the hot steam expands in the piston while losing pressure and losing temperature. Then the mechanism of the engine compresses the cooler steam in the piston to the same volume, temperature, and pressure that it started with. Then the next cycle begins.

Carnot analyzed this cyclical operation and discovered that it can be broken down into four, instead of two, distinct phases. Both the expansion and the contraction phases can be, *theoretically*, broken down into two parts. During the first part, the gas stays at the same temperature; during the second part, no heat transfer takes place. When combining the two parts we see that both temperature change and heat transfer takes place, as we said above.

This idealized theoretical operation allowed Carnot and others to mathematically analyze the operation of a steam engine and find out the limit (if any) of its efficiency. What is the maximum work we can extract from an idealized steam engine, by converting heat to mechanical energy?

The four phases of the Carnot-cycle are the following:

Isothermal expansion (no change in temperature) of the gas in the cylinder against outside pressure, at temperature T_1, increasing its volume from V_A to V_B, while its pressure drops from p_A to p_B. During this expansion, the gas extracts heat from the boiler and performs work.

Adiabatic expansion (no heat transferred, due to perfect insulation) increases the volume from V_B to V_C while its pressure drops from p_B to p_C and its temperature drops from T_1 to T_2. Since there is no heat transfer, according to the first law of thermodynamics, all the work performed by the gas decreases its internal energy.

Isothermal compression of the gas in the cylinder, at temperature T_2, reduces its volume from V_C to V_D , while its pressure increases from p_C to p_D. During this compression, the gas transfers $Q_2 > 0$ heat into the cooler and performs negative work (that is work has to be done on it).

Adiabatic compression decreases the volume from V_D to V_A while its pressure increases from p_D to p_A (we return to the starting point of the cycle) and its temperature increases from T_2 to T_1. Since there is no heat transfer, according to the first law of thermodynamics, all the work performed on the gas increases its internal energy.

With this theoretical breakdown of the operation, it is possible to analyze every one of the four phases and arrive at the maximum efficiency formula.

Sadi Carnot still believed the then accepted 'caloric' theory, which considered heat as a fluid. In spite of that, Carnot produced a correct description of the thermodynamic process of converting heat to mechanical work by a heat engine. From his work, Clausius was eventually able to deduce the second law of thermodynamics.

The Concept of Energy

The concept of energy is common knowledge. We use phrases like "I have no energy" or "he is 'energetic". But do we really know what energy is? People did not always use the word to describe what we mean by it today. In Newton's time, the main concept was 'force' as the agent responsible for movement and work.

Leibnitz observed in 1695 that the quantity of mass times the square of velocity (mv^2) is always conserved in interactions between parts of an isolated system. He called it "*vis viva*" or "living force" to describe what we call now the 'kinetic energy' of a system. In 1807, Thomas Young was possibly the first to use the term "energy" instead of *vis viva*, in its modern sense. However, even this earliest usage referred strictly to mechanical energy.

It took centuries to realize that there is 'something' in nature that can take many forms, transform from one form to another without diminishing in the process, and is the universal agent of change, movement, and work. We still don't know what this ghostly entity, 'energy', is - what it is made of, what it looks like. Richard Feynman gave us the most beautiful answer to these perplexing questions in *Lecture on Physics (Vol. 1; 4:1)*: "There is a certain quantity, which we will call energy, that does not change in the manifold changes which nature undergoes. That is a most abstract idea because it is a mathematical principle; it says that there is a numerical quantity that does not change when something happens….. It is important to realize that in physics today, we have no knowledge of what energy is."

The concept was arrived at when physics went beyond mechanics and started seriously studying electricity, heat, and chemical interactions. Half a dozen geniuses are responsible for developing the concept to what it means to us today.

James Prescott Joule (1818-1889)

When Joule started, the landscape of physics was confusing beyond description. This is often the case when we have a large amount of accumulated data in a field of study, without a unifying theory. Conflicting theories circulate, each advanced by some established authority and supported by its adherents.

It happened with electricity and magnetism when mainstream scientists still promoted the idea of 'electric fluids' and 'magnetic fluids'. When Faraday came up with the concept of electric and magnetic fields, he encountered adamant opposition until Maxwell unified all the theories with his Electromagnetic equations.

When Joule started his investigations, most scientists believed the Caloric Hypothesis, based on the assumed substance 'caloric', a 'chemical element' responsible for heat. Well-established scientists Black and Lavoisier supported this theory because heat appeared to be conserved in a thermally isolated system. They already knew about the conservation of mass and it seemed that they were on the right track. Lavoisier went as far as listing both 'caloric' and 'light' in his table of known chemical elements. According to the hypothesis, caloric could not be created or destroyed; its amount is fixed in the universe, but it can freely 'flow' from one object to another, resulting in a change of temperature that we can measure.

Rumford seriously challenged this concept when he demonstrated that unlimited heat could be produced by mechanical friction. If all that heat had been inside a metal test subject in the first place, it would have been molten.

However, Sadi Carnot, who also believed the caloric theory and used it in his calculations, nevertheless arrived at correct results. This was the landscape Joule entered.

So much was already known about mechanical motion, electricity, magnetism, chemistry, and heat that physicists knew that these different effects were somehow connected. A few of the known facts pointing to this connection were:

- Burning gunpowder in a gun barrel (chemical) resulted in (mechanical) motion of the bullet.
- Dissolving zinc (chemical) in a battery resulted in an electric current.
- Mechanical motion (friction) resulted in heat
- The electric current in wires produced heat.
- An electric current inside an electric motor produced motion.

But there was no explanation as yet of the relationship of these phenomena.

Joule was not a trained scientist. The second child of a well-to-do beer merchant in Salford, England, he was home-schooled and tutored until the age of 16, then took private lessons from John Dalton, the father of modern chemistry, and from John Davies, another recognized expert in Chemistry.

Having no systematic education at a university, as was the case with many of the early scientists - Volta, Ampere, Faraday, etc. – is a double-edged sword. On one hand, it gives the student maximum flexibility to pursue his own interests and allows him to strike out in new directions, sometimes unheard-of in established scientific circles. On the other, it results in debilitating 'holes' in their education, which slow their progress and cause costly detours. The lack of academic credentials often makes t difficult to be accepted by the scientific community.

Joule wasn't concerned by any of these handicaps – he was busy with his life in a protected and stimulating environment. He had his own laboratory in his father's house, where he pursued several experiments, mostly in chemistry and electricity.

Joule started out by studying electrical motors, trying to measure how much dissolved zinc in the battery resulted in how much lifting power. After many detailed and thorough experiments, he concluded that:

"...the duty of the best Cornish steam engine is about 1,500,000 lbs raised to the height of 1 foot by the combustion of one pound of coal, which is nearly equal to 5 times the extreme duty that I was able to obtain from my electro-magnetic engine by the consumption of a lb. of zinc [in the battery]. This consumption is so unfavorable that I confess I almost despair of the success of electro-magnetic attractions as an economical source of power..."

He was 22 years old when he wrote this, after 3 years of intensive experimentation and the invention of extra precise measuring instruments, such as his galvanometer. What he realized was that chemical action resulted in electrical action and electrical action resulted in mechanical action and it happened in a specific ratio, suggesting that there had to be definite equivalences between chemical, electrical, and mechanical processes. Following on this insight, he systematically studied eight different interactions and measured the quantities involved:

- chemical to thermal, electrical, and mechanical effects
- electrical to thermal, chemical, and mechanical effects
- mechanical to thermal and electrical effects

Up to this point, his main concern had been replacing steam engines with electrical engines, which could be cleaner and cheaper than coal. However, after the results of his experiments, he became excited by the philosophical implication of a suddenly glimpsed larger picture: a possible connection among all of the known phenomena studied by physics.

He attempted to publish his results in the *Philosophical Transactions* of the Royal Society but he, being an outsider to established scientific circles, and far in advance of current investigations, the Society printed only an excerpt of his paper *On the Production of Heat by Voltaic Electricity*.

This rejection never daunted his spirits and he proceeded with more experiments to explore the implications of his insight. Further pioneering results came one after the other. For example, Joule's First Law states that the heat produced by an electric current flowing through a conducting wire is proportional to the square of the current multiplied by the electrical resistance of the wire. Many more experiments - and results - followed.

Best known to the public is his crucial paddle-wheel experiment. He suspended a weight next to a volume of water inside a calorimeter. The descending weight powered an eight-blade paddle wheel inside the vessel to churn up the water and generate heat. From the mass of the weight, the distance of its descent, and the temperature rise inside the calorimeter, he determined the mechanical equivalent of heat.

The mechanical energy needed to raise 1 pound of water by 1-degree Fahrenheit, was equivalent to that needed to raise a 1-pound weight by about 800 feet. All his other experiments confirmed this conclusion. When he described it in 1847 at an Oxford meeting of the *British Association for the Advancement of Science*, his obscurity finally ended, due to the support of one man in the audience: William Thomson (later Lord Kelvin), professor of Natural Philosophy at Glasgow University.

After several significant recognitions from European scientific journals and institutions, Joule was finally honored at home: in 1850 he became "Fellow of the Royal Society" in England. The unit of energy is named after him: one *joule* is the energy expended (or work done) in applying a force of one *newton* through a distance of one meter. With this unit in place, the currently recognized conversion factor between mechanical work and heat is:

$$1 \text{ cal} = 4.184 \text{ J}$$

Joule demonstrated that mechanical work converts into heat at a specific rate. He also suggested that the reverse is true: heat converts into mechanical work at the exact same rate. However, none of his experiments proved this assertion directly.

Joule's accomplishments provided a major step toward recognizing the First Law of Thermodynamics (the conservation of energy principle) but further steps were required by other physicists before this universal law could be firmly established in physics.

William Thomson, Lord Kelvin (1824-1907)

Thomson's interest in thermodynamics started soon after he graduated from Cambridge. In 1846 he went to France, where he worked with prominent scientists and mathematicians. When he read Clapeyron's paper on Sadi Carnot's theory on heat engines, he became fascinated to the point of spending days and days searching for the original Carnot document in Paris bookstores.

He doubted Carnot's belief of heat being a fluid but was convinced that it must be conserved, like mass. He assumed that heat was not used up in a process, as Joule claimed by his heat-to-mechanical work conversion theory. He found Carnot's analogy to the work gained by water falling to a lower level, without disappearing, very powerful. As long as he saw Joule's and Carnot's theories as contradictory, he was tempted to accept Carnot as correct.

His attitude changed when he realized that Carnot's theory worked even without the assumption of heat as an indestructible quantity and found a compromise that satisfied him completely. Even though he still doubted the heat-to-mechanical work conversion, believing that heat is conserved, he stated: "The thermal agency by which mechanical effects may be obtained is the transference of heat from one body to another at a lower temperature".

But he was troubled by one thing:

"When thermal agency is thus spent in conducting heat through a solid [like cooling, without producing work], what becomes of the mechanical effect which it might produce? Nothing can be lost in the operation of nature – no energy can be destroyed".

Here it is, the first use of the word "*energy*" in the correct context. This concern finally won out over Carnot's assumption of heat being conserved. If heat can move from a higher temperature body to a lower one, sometimes with work produced in the process, sometimes without, then a third entity had to exist to account for the discrepancy. And this third entity, which he decided to call 'intrinsic energy', had to be the one conserved.

He still did not know what this 'energy' was, but he could quantify the internal energy change by measuring the heat in/out of the system and the work done by/to the system.

Just as Clausius did later, he mathematically defined energy as a **state variable**, dependent only on the volume and temperature of the system, regardless of how the system arrived at those values. He stipulated that interactions of any system with the surrounding environment consist of a combination of heating/cooling and work by/against the system. In this interaction, both the amount of heat contained in the system and the work performed by/on the system could change, but the overall energy in the universe remains constant.

This, in itself, was the discovery of the First Law of Thermodynamics: the total energy content of a system changes only by the heat put into it (positive or negative) and the work done on it (positive or negative).

Thomson published his theory in a paper titled: *On the Dynamical Theory of Heat* in 1851. He first used the term "*mechanical energy*" but changed it to "*intrinsic energy*" in 1856, when he realized that energy does not necessarily manifest in mechanical form but can appear in many different interactions.

So the concept of 'energy' and the law of its conservation was born. He did not call it the "First Law of Thermodynamics", but, in essence, that's what it was.

Thomson was not the only one who was convinced about the conservation of energy principle. In 1847 Hermann Hemholtz, a German physician and physicist postulated a relationship between mechanics, heat, light, electricity, and magnetism by treating them all as manifestations of a single 'force'.

Thomson made another substantial contribution to thermodynamics: interpreting the Gay-Lussac Law in 1848. He suggested that it was not the volume but the energy, of a gas that would contract to zero at –273 degrees Celsius. He proposed a new scale for measuring temperature, without negative values, where the zero point would correspond to –273 Celsius degrees, which that he called the "Absolute Zero Temperature". The rest of the scale would remain Celsius degrees shifted down, so that water would freeze at +273 Kelvin degrees and boil at +373.

He also developed the theory of "Energy Dissipation", which explained what happens to the energy that does not convert into work in heat engines. In 1852 he published a paper that explained how energy can be 'wasted' - that is, not converted to useful work - in the form of heat 'dissipating' into the environment in real-life non-reversible heat engines. This theory of 'energy dissipation' was a direct clue for Clausius who invented the concept of Entropy and formulated the "Second Law of Thermodynamics".

Thomson was one of these multi-talented prodigies: mathematician, physicist, engineer, inventor, teacher, and successful businessman. It seems unfair to the rest of us who hang on to one or two talents and hope to be recognized, in however small a way. He was knighted in 1892 for his contribution to the laying and successful operation of the first transatlantic telegraph cable between Ireland and Newfoundland. Without his assistance in design and instrumentation, the cable would not have worked even after 10 years of intensive effort on the project.

Rudolf Clausius (1822-1888)

Very little is known about Clausius's personal life. We know his educational records, his academic appointments, and his scientific papers, but no amusing story or anecdote seems to be circulating about his life. For some reason no biographer was tempted to write his story, possibly because he was a difficult person, arguing with almost everybody about scientific credit and the priority of inventions.

Clausius was a German patriot who earned the Iron Cross for being wounded in the Franco-Prussian war. He had a bitter dispute with Thomson over one of Joule's results that Thomson had quoted in one of his papers. Clausius insisted that a German had been the first to establish the result, not the Englishman Joule. Another dispute was with Peter Guthrie Tait over who was the first to propose the equivalence of work and heat. Clausius published an article in 1868 stating that not only did von Mayer have priority but so did the German nation. In 1872 when Maxwell published *Theory of Heat,* Clausius stated that the British were trying to claim more credit than they deserved for the theory of heat which, Clausius said, he alone was the discoverer.

What we do know from his scientific writings makes us admire him as a pivotal character in the development of Thermodynamics. The three major concepts: 'energy', 'entropy', and 'absolute zero temperature' were studied by the main contributors to the science, but Clausius was the first to express their relationships in clear mathematical form.

In theoretical physics, you have verbal suggestions and musings, insights, speculation about possible connections, and ideas about future research. And then you have precise mathematical definitions and equations that can be worked on by mathematical tools to investigate their validity and consequences.

The three concepts were touched on by Rumford, Carnot, Joule, Thomson, and even an obscure medical doctor, Robert Mayer, who expressed the concept of energy (calling it 'force') and its conservation principle as early as 1841.

Clausius was the one who combined all the known facts in clear mathematical notation, using the differential equations that physicists use today. In addition to connecting all the known dots, he came up with the new concept of 'Entropy' and the rules that govern its behavior. This led to what is now known as the "Second Law of Thermodynamics".

In essence: it can never happen that a cup of lukewarm coffee extracts heat from a cold table and heats up, while the tabletop becomes colder. Or, as Clausius stated in his 1854 paper: "*Heat can never pass from a colder to a warmer body without some other change connected therewith occurring at the same time*". The second law of thermodynamics seems so obvious when first encountered that one wonders why it was necessary to elevate 'trivia' to the lofty height of a basic law in physics.

To be sure, the first law of thermodynamics would not be violated, because no new energy would be created; it would only flow from a colder place to a warmer one. Yet, this has never been observed to happen. If heat could migrate, without external help, from colder to warmer places, then in essence we would have an engine that extracts heat from a heat reservoir, such as the ocean, and converts it into mechanical work. It would become a "*perpetual motion machine of the second order*" as Friedrich Wilhelm Ostwald (Russian-German physical chemist) called it. (A perpetual motion machine of the *first* kind would operate against, or violate, the law of conservation of energy). The human experience accumulated over millennia suggests that such a machine is not possible.

Clausius, being an accomplished theoretical physicist, proceeded to turn the law into a precise mathematical form. For this purpose, he had to invent a new concept, a new physical quantity that he called *entropy* from a Greek word for "transformation".

And now, in addition to the ghostly concept of 'Energy,' we have added another ghostly concept we call 'Entropy'. We have an intuitive grasp of the first: something exists in material objects that can change its form among the many observed manifestations such as mechanical, electrical, chemical, heat, nuclear, and then it can do some useful work for us. We still don't know what it is, but we have a personal relationship with it because our survival depends on it in an obvious way.

However, we know nothing, at this point, about what entropy is or how it relates to energy and to the other familiar physical parameters, like temperature, heat, pressure, and volume, for example.

Clausius arrived at the concept of entropy in a backward kind of way: mathematically, rather than conceptually. Even after he derived his equation and the properties of this mathematical entity, he still had no idea what it represented in a conceptual way. The physical meaning arrived later, with the help of other physicists who interpreted the concept and explained its meaning to us.

Back to Clausius: what did he actually discover? He was aware of Joule's discovery of heat to mechanical work conversion. He was also aware of Carnot's formula for the maximum efficiency of this conversion, which could be 100% only if the cooling reservoir was at absolute zero temperature. What he wanted to know was the precise mathematical formulation of this convertible energy.

He knew that the internal energy of thermodynamic systems could not all be completely released by any physical transformation into mechanical work, because during any transformation some of the energy was dissipated into the environment and some of it was used up internally by the molecules of the transforming medium colliding with each other.

As a model, he used Carnot's idealized 'reversible' engine, in which no heat is lost to friction or due to poor insulation. In this engine the ideal gas used as a medium was moving from one 'state' to another, uniquely described by the 'state variables' of volume (V), pressure (p), and temperature (T). During this transformation, there was always heat transfer from or to the steam inside the engine, and the small increments of this heat transfer (dQ) were measurable.

He discovered that summing up the dQ/T quantity, along any path taken in the cycle from status A (p_A, V_A, T_A) to status B (p_B, V_B, T_B) will give us the same value. <u>The value of the sum does not depend on the actual path taken from A to B</u>

This means that there is another quantity: S = Q/T that is a state variable, just as energy is: its value depends on the volume and temperature (or volume and pressure) of the system, regardless of how those values were acquired.

Due to the above, Clausius defined the entropy value S(P) for every point in the P(V,T) continuum of the studied process, such that the entropy of the system at that point is the <u>cumulative sum</u> of all the dQ/T values, obtained by transforming the system from absolute zero temperature value to the temperature T at point P.

$$S(P) = \text{(rev)} \int_{P0}^{P} dQ/T$$

where (rev) refers to the reversible process and P_0 is the status at absolute zero temperature where we assume that

$$S(P_0) = 0.$$

The symbol $\int$ stands for the 'cumulative sum' as explained in the "Your Math Toolkit" chapter.

      The entropy function defined this way always belongs to one particular system, such as a heat engine, that can move along the (V,T) continuum, as it changes its status from one point (V_1,T_1) to another (V_2,T_2).

      Finally, Clausius showed that any closed cycle can be approximated by a number of very small reversible and irreversible steps, and integrating (summing up) these steps gives us the following:

$$\oint dQ/T \ <= 0$$

where *the equal sign is valid only for reversible processes.*

      Until we get deeper into molecular and atomic physics, entropy is a mathematical concept, obeying definite rules and participating in definite relationships with the system under study.

      Clausius came up with the equations suggesting the master differential equation, familiar to us today, that connects energy, entropy, pressure, and volume and is valid for any system that goes through thermodynamic changes, without chemical changes in the process:

$$dU = TdS - PdV$$

We know that PdV represents the work done by the system against external pressure and dU represents the total internal energy change, so TdS will be the internally used up energy that can not be converted to external useful work. In other words: the momentary energy change at temperature 'T' is made up of the energy increase 'TdS' minus the energy decrease 'PdV' due to external work. This was the answer to Clausius's original quest to find a mathematical formula for the energy loss in heat-to-work conversion. *In this sense, entropy represents the part of internal energy unavailable for useful external work.* The unit of entropy is calorie per degree: cal deg^{-1}

Importance and application of Thermodynamics

Thermodynamics is the most abstract chapter of physics so far in our studies. Both Mechanics and Electrodynamics are familiar terrains – we all have seen moving bodies and electrical motors. We all have electrical outlets in our homes, and we have seen the stars and the planets – in physics we learn the details of familiar topics. Thermodynamics is different: we have to learn to deal with an *abstract* concept of a _system_ that can be virtually anything with a boundary that delineates it from its environment.

In thermodynamics, we want to know the rules of the non-chemical interaction (primarily work and heat exchange) between this system and its environment. The system can be our car's engine or vats in the food industry; in a factory manufacturing synthetic material, in a pharmaceutical plant compounding drugs, a biological lab producing DNA sequences, or a nuclear power station generating electricity. The range of applications is virtually unlimited. The implications for other sciences and for technology are far-reaching.

Most people are not concerned with all these technical details; we are happy that technology works. We have our cars, our safe food supplies, our medications, our electricity, our plastic toys, and our appliances. What do we care about the principles on which their manufacture depends? The curious among us do care. When my life depends on something, I like to know, at least in principle, what that something is, how it functions, what its rules are.

Thermodynamics will give me these answers and it will expand my vision and stretch my mind further than anything I have learned so far about the natural world. If I can master this, then I will be ready to dive into the world of atoms, the fundamental building blocks of the universe, and the often-incomprehensible laws that govern them. This is the final frontiers of the human mind, at least as far as the universe of the really small is concerned.

BOOK II - Modern Physics

Introduction

Physicists can't leave well-enough alone. At the end of the 19th century, we already had the scientific tools that enabled us to build clever machines, efficient transportation, and communication. We had electricity, we had the radio and telephone - the comfort in our lives surpassed anything previously imagined

Einstein's relativity theory sprung on the unsuspecting world was a warning of what was to come. With that theory we departed from the safe and predictable course that physics had been following for centuries, discovering phenomena and laws that made sense to us. Now we opened Pandora's box and started looking at things that we couldn't see - inside the atom. Once we peeked, physics descended through the rabbit hole into Alice's universe.

The history of Physics can be broken down to roughly three main phases. During the first phase, experiments were performed, phenomena discovered, laws established, predictions made, and applications found.

During the second phase, assumptions and hypotheses were advanced about the hidden causes of the observed phenomena, and then mathematical theories were constructed based on these assumptions. The predictions of the mathematical models were tested and found to agree with reality, however, nobody knew for sure whether the chosen assumptions were the right ones – they did not make any sense to the human mind.

The third phase is pure speculation, based on assumptions that could never be verified. Very sophisticated mathematical models were created, the predictions of which could never be tested, and no conceivable application could be seen forthcoming, though some Physicists see the proof of their theories in the aesthetic beauty of their equations. A word of caution: many theories in science, like Ptolemy's epicycles, were considered beautiful and worked for a thousand years, in spite of being utter nonsense.

Dealing with the second and third phases is a real challenge to the writer: how to describe the discoveries without following the beaten path of using evocative analogies. After mulling over this question at length, I decided to concentrate on two aspects of modern physics: the experimental results that are the bedrock of science, and on the applications that this science made possible. Both are real, both are with us every day of our lives, so we are on solid ground dealing with this part of reality.

The third aspect of this science is the assumptions and the interpretations physicists attach to the discovered phenomena. I will mention these, of course, but you have to take it with a grain of salt, never forgetting what we know for sure and what we only imagine.

Theory of Relativity - Start of the Weird

After Newton's dramatic breakthrough things settled down, many details were worked out, methods of calculation established, applications found.

The eighteenth-century whizzed by and nothing really dramatic – or fundamental – happened until electromagnetism was discovered by Oersted in 1820. Even after that, the major breakthroughs by Faraday, Maxwell, and Hertz only appealed to the scientists who could get excited by such esoteric topic as the unification of electricity and magnetism under Maxwell's laws of electrodynamics.

Even though practical inventions like generators, motors, electric lighting, telegraphs, the telephone, the radio, etc., were almost daily added to civilized life, still nothing extraordinary happened until Einstein formulated his Special Theory of Relativity in 1905.

Humanity has to deal with two kinds of knowledge: the practical and the spiritual. The practical is always welcome because it adds to our comfort and security, but we tend to take these in stride. The spiritual kind, on the other hand, excites because we gain a glimpse into something bigger than ourselves; something that suggests whole new worlds we have never before visited. Like Columbus's discovery of America – the adventure of the unknown and the unsuspected gives us the kind of thrill humans thrive on.

Einstein showed us that human beings had been looking at reality through the filter of our limited experience in space, time, and speed. Had we been able to travel close to the speed of light (300,000 km/sec) then we would have realized that our 'common-sense' comprehension of space and time are artificial and totally inaccurate descriptions of reality. Once that knowledge became widely known, our philosophical worldview was deeply affected.

Imagine, for a moment, that you were born deaf and color-blind. You have never heard sounds; have never seen anything other than black, white, and grey. The concept of sound is beyond your limits: you have never experienced it, could not imagine it, Beethoven is only a name encountered in books. The colour of a sunset or a Van Gogh painting is an abstract idea. And then, suddenly, miraculously, you start hearing sounds, seeing colors. You would feel like a superman, with incredible new powers of perception, totally beyond anything you could have imagined. Well, that is how we would feel if we could travel at relativistic speeds: the world we grew up in and learned to cope with suddenly expanded into a new dimension.

The implication of the Relativity Theory is a little bit like what space travel promised to be: a chance to go beyond the mundane, the well understood, the often-depressing predictability of the human condition. The question, "What's out there?" helps us forget about our daily struggles and fears on this planet.

Albert Einstein (1879-1955)

It is difficult to separate the person, the myth, and the scientist. In the popular mind, he is the saintly figure with unruly white hair, a gentle smile, and almost superhuman intelligence. "You don't need to be an Einstein to…" came to mean comparison with the utmost intelligence ever attained by humanity.

So many other, less worthy characters from politics, business, war, entertainment, sport, etc. have also been worshipped that a scientist finally elevated to a pedestal by enthusiastic crowds could only help science. However, turning a man into a myth has a negative side-effect. We somehow assume that Einstein was unique, beyond the grasp of any other human. His accomplishments became, almost by definition, something to hold in awe, not something to strive for. After all, who would have the nerve to compare himself, even in potential, to Einstein?

The myth is perpetuated by science writers and popularizers. It adds to the drama of narration if we emphasize extremes in a biography. No biography of Einstein I have ever read failed to refer to him as the "26-year-old patent clerk" who shook the foundations of physics.

The word "clerk" suggests an unskilled paper-pusher in a dusty office, filing documents, indexing catalogs. We don't expect clerks to have an education beyond high school, we don't expect them to have skills beyond those required for paperwork.

Einstein's job at the Swiss patent office required him to have a university degree and his assignment was to evaluate inventions and decide whether they were functional, original, and significant enough to warrant granting a patent. He needed a sound education in science in order to do that.

Many other attributes were added to the myth to increase the drama. Almost no one can resist repeating that Einstein was slow in school, that his teachers thought he wouldn't amount to anything, and, even in university, he was called a "lazy dog". The contrast between the almost pathetic start in life and the God-like stature he later acquired is so delicious that we just can't let it go. Einstein himself disliked the adulation as characterized by one of his quotes: "To punish me for my contempt for authority, fate made me an authority myself".

Einstein was a true genius, on par with Newton and Maxwell. However, just as those two did, he acted more like a catalyst and a synthesizer than an inventor of completely new theories that had never occurred to anyone before.

We know that Newton's idea of universal gravitation was suggested by Robert Hooke in a letter to Newton and we know that Maxwell's equations are built on facts and principles already well known. What both Newton and Maxwell did was synthesize all the known facts into coherent theories that explained the accumulated data and provided a framework, integrated both mathematically and conceptually, that others could, in turn, build on.

Many of the ideas that are fundamental in Einstein's Relativity Theory had been suggested before, in one form or another, by contemporaries such as Poincare and Lorentz. However, it was Einstein who brought clarity and order into the confusion and fitted the pieces together for the first time. I am certain that, had Einstein never existed, someone else would have made the same discoveries.

There is controversy about Einstein's knowledge of the crucial Michelson-Morley experiment that is still today considered as the first undeniable evidence against the ether assumption. Einstein seems to have contradicted himself in public statements about this controversy.

However, all of the above are truly irrelevant when evaluating Einstein's legacy and contribution to science. I love and admire Einstein and, for this reason, I want him to be known for what he really was. He has the right to be appreciated for his accomplishments, rather than for the dramatized and exaggerated image many have about him.

Relativity theory

Relativity theory deals with moving systems, observers, and laws of nature. The moving system can be a planet, a spaceship, a train, or a ship crossing the ocean.

On top of system 'A' there is an observer sitting at his workbench and doing experiments in mechanics, electrodynamics, etc. He observes the results of his experiments and deduces the laws of nature governing the behavior of material objects and processes. On top of another system 'B', which is moving relative to system 'A' at uniform speed, in a straight line, another observer is doing the same.

For Einstein, it was inconceivable that the laws of nature would appear different to the two observers doing their experiments. If they play billiard in their system, the billiard balls behave the same way in both systems. If they connect a wire to a battery and run an electric current through it, the magnetic field created around the wire will be exactly the same. The relative uniform speed of the two systems should not affect the outcome of the experiments in either system.

Uniform speed is important because if you allow one system to accelerate (charge speed and/or direction) then the behavior of the billiard balls in the accelerating system would definitely be different.

The famous paper

Einstein published it in 1905 under the title *On the Electrodynamics of Moving Bodies* and it starts: "It is known that Maxwell's electrodynamics - as usually understood at the present time - when applied to moving bodies, leads to asymmetries which do not appear to be inherent in the phenomena. Take, for example, the reciprocal electrodynamic action of a magnet and a conductor. The observable phenomenon here depends only on the relative motion of the conductor and the magnet."

On the face of it, the two experiments seem quite different. In one case, I have the loop of wire on my workbench and wave a bar magnet over it. In the other case, I have the magnet lying still on my workbench and move the wire over it. This is how Faraday saw it and described the two experimental results in two different ways. The second Maxwell equation, which is Faraday's Law, describes the induced electric field as the function of the changing magnetic field. The *'changing magnetic field'* assumes an observer for whom the field is changing.

If the observer moved with the magnetic field (uniformly drifting eastbound, for example), then the field would not change for him. Does this mean that no electric field is generated by the same magnetic field? How could an 'objective' physical phenomenon depend on where we observe it from? If something is happening in nature, it should be happening whether we watch it or not, never mind the location of our workbench.

What Einstein made us realize is that the *exact same thing is happening in both cases*: the loop and the magnet are moving _relative_ to each other. How we see them depends entirely on the vantage point from which we observe the experiments. If we place our chair on top of the magnet, then it is the loop that moves. If we sit on the loop, then the magnet moves. If we jump up and down between the two, then both the magnet and the wire move. Einstein's great breakthrough consisted in realizing that an entirely physical process, the actual physical thing happening, should be completely impervious to the location in space from which we observe it.

Maxwell's equations contain the constant 'c' which was measured by both Maxwell and Weber-Kohlrausch as the speed of light in a vacuum. The speed of light, relative to what? - Einstein asked. The general assumption at the time held that light was transmitted through a hypothetical medium, called the luminiferous aether; however, all experimental attempts to locate and describe this aether failed. Einstein then proceeds:

"… the same laws of electrodynamics and optics will be valid for all frames of reference for which the equations of mechanics hold good. We will raise this conjecture (the purport of which will hereafter be called the "Principle of Relativity") to the status of a postulate, and also introduce another postulate, which is only apparently irreconcilable with the former, namely, that light is always propagated in empty space with a definite velocity c which is independent of the state of motion of the emitting body."

Here is the heart of the special relativity theory: the famous two postulates:

- The same laws of electrodynamics and optics will be valid for all frames of reference for which the equations of mechanics hold good.

- Light is always propagated in empty space with a definite velocity 'c' which is independent of the state of motion of the emitting body.

The consequences of these postulates are as fantastic as they are inevitable. What Einstein gained from these assumptions was clarity and consistency of the theories and experimental data collected in physics for the last 300 years. What he gave up was a common-sense view of the universe.

The audacity of his suggestion reminds me of a joke I heard a long time ago: A husband arriving home unexpectedly finds his wife in bed with another man. When he confronts her, the wife denies her infidelity and asks indignantly: "Who do you believe, Elmer, me or your eyes?"

I don't know how Einstein would have replied if he were put to the test, but in the area of physics, he was prepared to disbelieve his eyes – not because he was willing to discard experimental data, but because he assumed that our lack of experience with great speeds (close to 300,000 km/sec) had not allowed us to observe the actual behavior of nature in such extreme circumstances.

The first postulate seems reasonable: if all the laws of mechanics work perfectly identically in two systems of observation, then why should electrodynamics (and optics) behave differently? We assume nature to be consistent and if one aspect of it is identical in two systems, then we expect all aspects to be identical.

The second postulate is the one that goes against common sense. As I will explain in the section called "Einstein's 'No'", the only way this is possible is if the measurements of space and time yield different results for observers moving at a uniform speed relative to each other, even if they use identical and synchronized instruments.

No wonder it required Einstein's exceptional insight and courage to put forward such a preposterous proposal. The rest of the paper deals with examining the consequences of these two postulates.

Galilean Transformation

We found the first expression of a relativity principle in both Galileo's and Newton's writings. What they believed is now called the "Galilean Relativity Principle". It says, in Newton's words:
"The motions of bodies included in a given space are the same among themselves, whether that space is at rest or moves uniformly forward in a straight line".

Galileo assumed, though he never actually performed the experiment, that if you dropped a rock from the top of a mast of a moving ship, it would land at the base of the mast, exactly as if the ship were a rest. This seems to make sense if we realize that the ship and the stone were moving forward at the same horizontal speed, therefore the vertical drop of the stone should not make any difference to its distance from the mast.

Now consider the following situation: a spaceship is leaving Earth toward the moon at the escape velocity of 11km/sec. The distance of the moon from Earth is 384,400 km. This distance never changes, but the distance from the spaceship to the moon changes from second to second. Assuming that the speed of the spaceship remains the same, the distance shrinks by 11km every second.

If we denote the spaceship's speed as 'v' km/sec and the moon-Earth distance with 'x_0', then the spaceship's distance to the moon 't' seconds after the departure will be:

$$x = x_0 - v t$$

Another way to express this relativity principle is via mathematical transformation formulas: Assume that two observers (one on Earth, the other in the spaceship) are moving relative to each other in a straight line, with uniform speed 'v'. Each observer measures distances on the same line, relative to its location in space, described by the well-known coordinate system. For the sake of simplicity, they choose two coordinate systems in such a way that the two 'x' axes are on the same line, pointing in the direction of the relative movement of the two systems. The other two pairs of axes are parallel pairs and that at t=t'=0 moment the origins of the two systems coincided.

In this setup, the x, y, z coordinates of a point in space (as measured in the first system), and the x', y', z' coordinates of the same point in space (as measured in the second system) are related to each other according to the following Galilean transformation formulas:

$$x' = x - v t$$
$$y' = y$$
$$z' = z$$
$$t' = t$$

where t' and t are the common time measured in both systems.

Newton's equations should work in both systems, with each observer using his own measurements. Since we just determined how these measurements are related, then, if we use these transformation formulas for the Newton equations, then, doing the calculations, we find that the equations are in the same form in both systems. That means that the mechanical laws are identical in both systems, just as the Galilean Relativity Principle demanded.

However, if we use these transformation formulas for the Maxwell equations, then the equations will be in different forms in the two systems. In other words: the laws of electrodynamics are not invariant in regards to the Galilean transformation formulas. This means that if I performed the same electrodynamic experiments in both systems, I would get different results.

Lorentz Transformation:

The question naturally arose: what kind of mathematical transformation formulas would leave the Maxwell equations in the same form? H.A. Lorentz was the first who answered this question. He proposed that the Maxwell equations stay in the same form if we us the following transformation formulas:

$$x' = \gamma (x - v t) \qquad \text{(i)}$$
$$y' = y \qquad \text{(ii)}$$
$$z' = z \qquad \text{(iii)}$$
$$t' = \gamma (t - x v /c^2) \qquad \text{(iv)}$$

where $\gamma = 1/(1 - v^2/c^2)^{1/2}$ $(1 < \gamma < \infty$ for $0 < v < c)$ and 'c' is the speed of light in vacuum.

These transformation formulas are known as the "Lorentz Transformations".

If we solve equations (i) to (iv) algebraically for x and t, then we will have the transformation formulas from the second to the first system as:

$$x = \gamma (x' + v t') \qquad \text{(i')}$$
$$y = y' \qquad \text{(ii')}$$
$$z = z' \qquad \text{(iii')}$$
$$t = \gamma (t' + x' v /c^2) \qquad \text{(iv')}$$

which looks exactly like the first set, considering that the relative speed of the first system to the second is '-v'. It would be really surprising if it came out differently!!!

So far, there was no new physics involved, only a mathematical trick of calculating backward from the Maxwell equations to see what transportation formulas would leave them in the same form. If we use these formulas, then the form of the Maxwell equations remains the same when we convert from one system to the other.

Conflict in Physics

There seemed to be a conflict here. If we choose the Galilean transformation, then the laws of mechanics remain the same, but the laws of electrodynamics do not. On the other hand, if we choose the Lorentz transformation, then the laws of electrodynamics remain the same, but the laws of mechanics do not. The only way to reconcile this conflict is to decide to correct either Newton's equations or Maxwell's equations. We can not leave both unchanged, because it is inconceivable that any natural law would behave differently in two systems moving at a uniform speed relative to each other.

Experimental data eventually convinced physicists to keep the Maxwell equations unchanged and modify Newton's equations. This modification is a simple one: all we have to do is introduce a correction factor into Newton's second law: "Force equals mass times acceleration". Using the definition of acceleration and speed, we will have:

$$\mathbf{F} = m\mathbf{a} = m \, d/dt \, \mathbf{v} = m \, d/dt \, (d/dt \, \mathbf{x}')$$

In another system moving relative to the stationary one, we have to use that system's measurement for x, according to the Lorentz transformation formula described above: $x = \gamma \, (x' + v \, t')$

Therefore, the correction factor needed is: $\gamma = 1 / (1 - v^2/c^2)^{1/2}$ so the new form of Newton's second law will become:

$$\mathbf{F} = m \, \mathbf{a} / (1 - v^2/c^2)^{1/2}$$

The physical meaning of this modification: the mass is dependent on the speed of the observer relative to the mass, according to the following formula:

$$m = m_0 / (1 - v^2/c^2)^{1/2}$$

where 'm_0' is the mass of the object measured by an observer who is at rest relative to the object, and 'm' is the mass measured by another observer moving at uniform speed 'v' relative to the object. If the observer is not moving relative to the mass, that is $v=0$, then $m=m_0$. That is why m_0 is the rest mass of the object.

Since $v<c$ (according to Einstein's postulate), $(1- v^2/c^2) < 1$, so we divide m_0 by a true fraction which results in $m > m_0$

The mass of an object measured by an observer in a system moving at a uniform speed relative to the object is greater than the mass measured by an observer at rest relative to it. With this modification in place, if we use the Lorentz Transformation formulas, then the laws of mechanics will also stay the same in both systems moving uniformly relative to each other.

Consequences of the Lorentz Transformation

The question is: what price do we have to pay for this consistency in our physics theories? What are the inevitable logical consequences of this decision? They turned out to be mind-bending, to say the least.

I had a conversation once with a physicist at the Perimeter Institute in Waterloo, Ontario. I asked him how he would explain the invariance of the space-time interval in relativity theory to laymen, without using mathematics. He scratched his head for a while and then proceeded to verbalize the equation using words instead of symbols. When I told him that, he laughed and admitted what we both knew: relativity theory is a mathematical construct, a direct consequence of Einstein's postulates.

With this in mind, I decided not to follow the usual path of verbalizing equations and/or using analogies but rather use algebra to show how the non-intuitive consequences of the postulates are inevitable by the rules of logic and mathematics. The brave and curious will find this in the "Next Level" section of this book.

I will only summarize these consequences here. All of the following will compare how two observers, moving relative to each other, at uniform speed, experience the world around them.

Relativity of simultaneity

Events that appear simultaneous to one observer will appear to happen with a time interval between them for the other.

Time synchronization

If an observer in system 'A' examines all the clocks at different locations in system 'B' which is moving at a uniform speed relative to system 'A', he finds that all the clocks in system 'B' are out of synch with each other, because their time depends on their location as measured in system 'A' according to the Lorentz transformation.

Time dilation

A time interval between two events (such as two lightning bolts hitting the ground) measured by an observer who is moving relative to the events is longer than the time interval measured by an observer who is at rest relative to the events. Therefore, the clocks in the moving system seem to slow down compared to the clocks in the local system.

Lorentz contraction

The length of an object measured in a system that is moving relative to it is shorter than its length measured in a system in which the object is at rest.

Relativity of mass

The mass of an object measured in a system that is moving relative to it at a uniform speed 'v', where m_0 is the mass measured in the system in which the mass is stationary:

$$m = m_0 / (1 - v^2/c^2)^{1/2}$$

At very low speeds (when v^2/c^2 is tiny) the $(1 - v^2/c^2)$ correction factor becomes so close to 1 that, for all practical purposes, we all measure the same mass for the same object, even if we move (uniformly) relative to each other. At high speeds, however, the mass of the same object will be measured as greater by the observer who is moving relative to the mass, compared to the value measured by the one who is at rest relative to it.

Equivalence of mass and energy

And this is the most famous formula in all of physics, one that everybody knows. There is an equivalence between energy and mass, according to:

$$E = m\,c^2$$

Where 'c' is the speed of light.

Speed addition

The speed addition formula is one of the most counter-intuitive aspects of the relativity theory; however, it has been proven over and over again with no exception.

If platform 'B' (such as a spaceship) is moving relative to another platform 'A' (for example, Earth) at speed v_1; and a bullet is fired from platform 'B', in the same direction, at speed v_2,(relative to 'B') then the speed of the bullet is measured from platform 'A' as:

$$v = (v_1 + v_2)/(1 + v_1\,v_2\,/\,c^2)$$

where 'c' is the speed of light in vacuum (300,000 km/s).

As we can clearly see, speeds don't add up: the closer they are to the speed of light, the more we lose when combining the two speeds. Applying these formulas to everyday life we will see why the discrepancy never showed up in experiments before. If I travel on a bicycle at 20 km/h and throw a ball at someone standing on the ground ahead of me at a speed relative to the bicycle of 5 km/h, then the ball will NOT hit the person with a speed of 25 km/h. The speed of the ball hitting the person will be slightly less than that, as given by the

$$(20 + 5)/(1 + 25 \times 5/c^2)$$

formula where 'c' is the speed of light (300,000 km/sec). The correction factor with which I divide the expected value will be about 1.000,000,000,000,000,010,717 (c=300,000km/s = 10.8 x 10^8 km/h; c^2=116.64 x 10^{16} ; 125/ c^2 = 1.071673525 x 10^{-16})

At such a low speed, this loss is so small that it is almost zero. Almost, but not quite. As we think in terms of faster and faster speeds, the correction factor becomes more and more significant. If my bicycle could go at <u>half</u> the speed of light and I could throw the ball also at <u>half</u> the speed of light, then the ball would hit the unfortunate person at (c/2 + c/2)/(1 + (c/2 x c/2) /c^2) = 4c/5; that is, four-fifths of the speed of light, instead of the full speed of light as we would have expected. I would have lost c/5 speed (which is 60,000 km/sec) somewhere - a significant loss. Finally, if both speeds were equal to the speed of light then adding the two velocities would yield:

$$(c + c)/(1 + (c\ c)/\ c^2) = 2c/ (1 + 1) = c.$$

Exactly the same, as if at that point I could not add even a tiny fraction to the speed of the ball (relative to the ground) by throwing the ball from it at the speed of light.

……………..

Now that we have looked at the mathematical consequences of relativity theory, I would like to bring something to your attention. Please look at the Lorentz Transformation formulas once more, as applied to space and time intervals ('x' is a space coordinate, 't' is a time coordinate, 'v' is the relative uniform speed of the two systems, and the symbol 'Δ' means a small difference):

$\Delta x' = \gamma\ (\Delta x - v\ \Delta t)$
$\Delta t' = \gamma\ (\Delta t - v\Delta x\ /c^2)$
where $\gamma = 1/(1 - v^2/c^2)^{\frac{1}{2}}$ $(1< \gamma< \infty$ for $0<v<c)$

You can't help seeing how space and time are intertwined. Two events (like 2 lightning bolts) happen at some space and time coordinates, measured independently by two observers, each in his own system. When one of the observers calculates how the observer in the other system would have measured the space and time coordinates of the same two events, he will find the following:

The distance between the two events measured in the other system depends not only on the distance measured in the local system but also on the time interval between the two events in the local system when the measurement was made.

The time interval between two events measured in the other system depends not only on the time interval measured in the local system but also on the difference of spatial coordinates in the local system where the measurement was made.

With this in mind, do not try to find 'paradoxes' by *imagining* how two observers could actually peek at each other's systems over huge distances while moving away from each other at close to the speed of light. Never forget the value of the speed of light: *it can circle our planet seven times in one second*! Our imagination is not up to coping with this number, because we don't have any experience in this region.

In the notorious "twin paradox" example, two twins are moving at a near-light uniform speed relative to each other. Based on the Lorentz Transformation, each is convinced that the 'other' one, speeding away from him, is younger, because his clocks slowed down. So, which of them is right? The answer is: they are both right in their assumptions, in their own systems, while both feel they are aging at their normal pace.

You might be tempted to ask: "but what is the truth in objective reality"? The answer is: *there is no objective reality in any practical sense* of the word, only relative reality.

Every statement anyone ever makes is based on his own personal experience in his own system of space and time. Not one of us is a god able to see all of it from 'above' and at the same time. *This is the hardest to understand for most people*. All we can say with certainty are the answers to the following questions:

- What is it we have observed?
- What are our assumptions about unobserved phenomena?

- What are the logical/mathematical consequences of these assumptions?
- Have we been able to observe these consequences?

The rest is pure speculation without supporting evidence and may or may not have anything to do with reality. This is the universe we are part of, whether we like it or not. We have no other option than what I outlined if we want to stick to what we <u>can know</u>.

As one practical example, consider the Large Hadron Collider at CERN in Switzerland. Two beams of protons are racing towards each other, at near the speed of light relative to the ground. Their speeds are a little above 0.999c

What is their speed relative to each other when they meet? Surprise, not 1.999c but a little above 0.999c. This is a practical application of relativity theory working today in an international multi-billion-dollar project.

Another example: 'muon' is an elementary particle produced in the upper atmosphere by cosmic rays. Its lifetime is 2.2 microseconds. It should never be able to reach the Earth's surface even at close to speed-of-light velocity. Yet, we can detect them at Earth's surface regularly. Why? Because of time dilation, the muon time slows down relative to us.

During the Manhattan Project, when physicists grumbled about quantum theory not making sense, they were told to "shut up and calculate!" Nothing illustrates more the dual nature of modern physics: it doesn't make sense and it does work!

Space-time

An excellent book published in 2003 by Richard Wolfson is *Simply Einstein*. It has a chapter dealing with the same issues, under the title of: "Is everything relative?"

We have already dealt with two absolutes: the laws of nature and the speed of light, which we expect to be the same for everyone, anywhere. These were the first two absolutes that relativity recognized. But there are more. The next absolute Einstein created for us is space-time. What is it? The name suggests some kind of relationship or interdependence between space and time. Let's explore what it means.

Newton considered space and time as separate absolutes, with no connection between the two. We measured space in its three dimensions with rulers. We measured time, flowing evenly for everyone, with clocks. This seems obvious in view of everyday experience in the world.

How did it change? What does it mean to us? How is it absolute? The best explanation I have ever read was given by Brian Greene, in his magnificent book: *The Fabric of the Cosmos*.

"We are used to the fact that objects can move through space, but there is another kind of motion that is equally important: objects also move through time. ……When you look at something like a parked car which from your viewpoint is stationary - not moving through space, that is – all of its motion is through time. …... But if the car speeds away, some of its motion through time is diverted into motion through space."

Thinking about relativity in this fashion actually makes me *feel* the four dimensions in which I exist, and I can see how the three dimensions of space and the one dimension of time are interwoven to make a complete environment and the notion of time slowing down the faster we go through space actually seems *inevitable!*

One aspect of all this needs to be emphasized. The above quote assumed an observer watching the moving object and describing it from his own point of view. The implication that the observer was stationary while the object was moving is a misleading one: the status of 'being at rest' is arbitrary, since there is no absolute space to which we could compare things. "Rest" will become a relative term: every observer is 'at rest' in his own system.

Because of this symmetry, the description of the exact same interaction between the object and the observer can be reversed, in which we consider the object being 'at rest' in the system attached to it and the observer moving away at some speed. In this case, it is also true that the previously observed object, will 'see' the clock attached to the previous observer running slower and slower as more and more of his speed through time is diverted to speed through space.

Yet, there is <u>no symmetry</u> in the so-called "twin paradox" because the traveler who returns much younger than everyone who was left behind did not spend all his time in an inertial system, moving at a uniform speed relative to the point of departure. He had to decelerate, stop, turn back, and after more positive and negative acceleration, make his way home. So, you can not apply time dilation symmetrically.

What is space-time?

Up till now, we thought of space as the totality of the expanse of the universe. Every place in a three-dimensional continuum belongs to it. Up till now, we thought of time as the totality of all past and future in the universe. Every moment in a one-dimensional continuum belongs to it. Both of our Newtonian space and time concept are static media that stretch out to infinity in each direction.

Space-time is more than a sum of the two. The combination of space and time becomes event-oriented: All events that ever happened, or will ever happen, in the universe, will become a four-dimensional point in the continuum we call space-time. Space-time is dynamic, seething with action, change, movement.

My own personal life is a four-dimensional line in space-time. Everything I have done and will do, anything that ever happened, and will happen in it, is on this line.

In some religions, God is described as omniscient – knowing all events in all times, past and future. What if there was a being who had the senses to perceive the totality of four-dimensional space-time and read it like a book?

In such a book, evocatively described by Brian Greene, each page would constitute a time-slice of the space-time continuum: all the events that happened in all of space in a particular moment of time, from the perspective of one observer. Another observer, moving uniformly relative to the first, would slice the 'space-time book' into different pages because two events that happened simultaneously for observer 'A' would be on the same page for him but, according to the relativity of simultaneity, they could be on different pages for observer 'B'.

I once read a very interesting science fiction story: a maverick scientist goes into business telling people the exact date of their deaths. He offers a million dollars to the first person who proves him wrong. He is never wrong – his predictions come true, one after the other, with unerring precision. He can follow anybody's four-dimensional lifeline through space-time into the future and see where it ends. The insurance companies don't like him at all, because people whose demise he had predicted for the near future take out large policies. They threaten him, but he keeps on doing it (I guess he did not like the insurance companies either). When his murdered body is found, in his diary they find the exact date and time of his own death pre-recorded in it.

This story assumes that the future is pre-determined, a concept disputed by modern quantum mechanics, but the jury is still out on that one. Even if the future is flexible, one's lifeline through space-time still exists, with the future part of it wiggling a bit as influenced by random events.

Philosophical implications - What does it mean?

Most of the books on relativity written for the general public or even for the student, deal with the reasons Einstein proposed this theory, the consequences of accepting the postulates, and the experimental proofs of the theory, not the philosophical implication. That's what I find most exciting.

We will never intuitively understand and accept relativity, because, due to our limitations in speed we lack personal experience of the relativistic world. Yet, we understand it intellectually, mathematically, and we would like somehow to relate to it emotionally. We know that it means something, but we cannot tell what. We can only speculate.

What kind of reality are we talking about when we accept the relative nature of space and time, mass and energy? What happened to the objective reality we assume to exist independently of our minds? The very foundation of science had been, until Einstein, the assumption that there is a world with its objective reality, its unchanging laws. All we had to do is find out how it was put together.

We fancied ourselves as the curious tinkerer who stumbles upon an intriguing machine and, by studying its workings and the visible parts of the mechanism, can understand its function and purpose. We could not approach this task without the assumption that the machine is always the same: what was true yesterday is still true today; what is true for me is also true for everyone else. These were the absolutes we expected from the subject of our investigation.

And then suddenly Einstein comes along and tells us that we were all wrong, there are no absolutes, what appears to one person may look completely different to another: it all depends on how and from where we look at things.

Or did Einstein say that? Did he abolish all the absolutes in our universe? Or did he abolish some, leave some alone and create new ones?

These are important questions and the answer lies in ruthless honesty. We have arrived at a point where we have no choice but to admit our limitations. We can try to cheat by using analogies and evocative images; try to draw a parallel between the reality we have been familiar within our daily lives and the reality we measure with instruments and describe with mathematics.

In my opinion, such shortcuts are confusing and pointless: *reality is what it is* and by attaching seductive imagery, we falsify it. On the other hand, if we accept that we have observed facts and logical consequences, without reaching for visual interpretation, then we are on solid ground and ready to follow into the totally obscure world of quantum physics.

We must **always** be aware: **what do we know, how do we know it and what are the limits to our knowledge**?

In the case of Relativity Theory, we know the following:

- No aether has ever been observed in myriad experiments performed in order to detect it.
- Newton's and Maxwell's theories are incompatible.
- Detailed and advanced experiments agree with Maxwell.
- Only the Lorentz Transformation leaves the Maxwell equations invariant among uniformly moving systems.
- Einstein's postulates are required for the Lorentz Transformation.
- The mathematical/logical consequences of these postulates are inevitable.
- Every prediction arrived at logically, based on these assumptions, has been experimentally verified without exception, over a hundred and nine years.

Maybe, if we could experience the four-dimensional space-time continuum directly, physically, we would have a solid foundation under our 'philosophical feet' again. However, at the moment, we can take pride in our honesty and courage in the following truth to where it leads.

How do we live with it?

I am often asked by my students: "What good is something that I can never experience? Isn't it cruel to show people a world they will never reach? Show them a world that goes way beyond their abilities to live in and, while as in Plato's cave, they are forever doomed to watch its mere shadows flicker on the walls?"

My answer is, invariably: It is important to understand reality, however strange and alien it appears. We need to know that we are not gods; that our job is not to fight, subdue and defeat nature, but to respect it with the same humility an ordinary leaf would respect the tree it belongs to if leaves were capable of such emotions.

We may not ever fully live in that world, but we can always sidle up to it, get closer and closer as we see more clearly. Maybe our great-grandchildren, whizzing about near the speed of light in their spaceships will not find time-dilation and length contraction as weird as we do. If you grow up with something, it becomes part of you, you take it for granted.

In the meantime, we can, and we do, use these strange laws in calculations when we increase our comfort and security with new technologies. We don't need to fully comprehend relativity in order to accept it and build it into our lives.

Many consider the 'gift of science' to be a mixed blessing. I have friends who refuse to own a computer, threw out the TV and VCR from homes that they built "off the grid", heat and cook with wood stoves, just like their grandfathers used to. I do understand their suspicion of technology, their mistrust of scientists and engineers. I am aware of the self-serving lies, the irresponsible rushing into applications, and the smug conceit and condescension of many. I do know of pollution, climate change, the extinction of countless species, depletion of resources that are all side effects of science and technology. I am also aware of the uncountable advantages science has given us – advantages we take for granted and will not hesitate to use when in need. Carl Sagan was troubled by this contradiction when, despite being an animal-rights advocate, when terminally ill he took advantage of medical techniques that could never have been developed without experimenting on animals.

The answer, as usual, is sought in the backswing of the pendulum. The answer, as usual, is wrong.

First, there never was a "Golden Age" that we can escape back to. In his magnificent book *The Third Chimpanzee,* Jared Diamond devotes a whole chapter to the topic of "The Golden Age that never was". Second, the problem is not science and technology, but the way we collectively use it and allow it to be used. We are all responsible. Third: we can't go back, can't unlearn what we know, can't forget our scientific heritage of millennia. Some would remember, some would tempt us with an irresistible bribe of living easier, healthier, longer. Soon we would be back where we are now.

So what are we to do?

My advice: Do the best you can and hope that enough sanity prevails to see us through this technological adolescence without destroying ourselves and a good chunk of the planet we live on. Maybe, at great cost to ourselves, we can learn how to use science in a responsible way, to the benefit of all, without destroying the world.

If we do, the rewards could be breathtaking. If we did not spend most of our energies on making weapons of mass destruction and mindless distraction for both the rich and the poor, then we would have the resources to create a real paradise.

We could easily have abundant, cheap, and clean energy for all if we took nuclear fusion research seriously. We have had practical nuclear fusion for 60 years in hydrogen bombs. We can liberate the energy; we 'just' need to learn how to control it in a safe and sustainable way. Once we had this energy source, so many things would automatically follow. We could stop using fossil fuels and thus eliminate the major cause of pollution and the savage imperialist wars now being waged over third-world oil reserves.

We wouldn't even have to abandon capitalism to solve the class hatred now tearing many nations apart. If no one was hungry, cold, living in fear and poverty, most people would be happy with their lot of comfort and security. Relatively very few people want to live in palaces and mansions and have private jets to fly about.

Once we solved our problems here on Earth, then we might take a new look at the stars, to see if we can find a way to visit them? Imagine the thrill of stepping off this "pale blue dot" to see what is out there? No 'reality show' could equal the awe and wonder of rising above our pitiful self-imposed limitations.

Since we can't go back to a "Golden Age" that never was, our only choices seem to be: self-destruction or utopia.
I vote for the second.

Atomic Physics - Science that kills

We all remember from science class: substances that can be broken down by chemical means are "**compounds**"; those that can not are called "**elements**". Most of us could name some elements: oxygen, hydrogen, iron, copper, gold, lead. Familiar compounds are: "water" "carbon monoxide", "benzene", "sulphuric acid", "sodium chloride" (salt). If hard-pressed, we can even recall that elements are made of a single type of atom (e.g. oxygen is made of oxygen atoms) and compounds are made of two or more elements combined into a molecule (e.g. a water molecule is made up of two hydrogen atoms attached to one of oxygen).

So far, so good.

We were taught these facts and we took them on *faith*. Even though we had never ever directly encountered an atom or a molecule of any kind, we had no reason to doubt our teachers. The story was reasonable, consistent; it made sense.

Very few of us, very rarely, think about those times not too long ago, when these 'facts' were not obvious at all; when intelligent, competent scientists rejected the "atomic hypothesis" out of hand.

What convinced them otherwise? What shall I tell my granddaughter, if she asks: Grandpa, how do you know that there are atoms when you can't see them?

The most important question every student should ask if he wants to understand quantum physics is: what is it that we know for sure about atoms? What fundamental experiments convinced us that there had to be such things?

When we get to the weird world of the very small, and we are told that atoms, electrons, and other building blocks of matter behave like solid objects in some cases and like waving energy packets in other cases, we will have to recall the fundamental experiments and see if we assumed too much, maybe more than the facts warranted.

Beware of what you were taught in elementary and high school! Almost everyone who is not a trained physical scientist will describe the atom as a miniature solar system, with the nucleus in the middle and the electrons 'orbiting' around it at different distances, just like the planets around the sun. However, if we read further, we are told that it does not make any sense to talk about the location and speed of an electron in its 'orbit'. Some textbooks will even tell us that there is no point talking about an electron being anywhere inside an atom until a measurement is made -- the electron will become 'real' only *at* that moment and only *for* that moment. These ideas led some physicists even to question the existence of reality, suggesting that there is no independent reality outside consciousness.

All of these things were suggested in many forms by completely sane, genius-caliber, highly competent scientists, for very good reasons.

My only argument with most textbooks on physics is that these confusing properties of the building blocks of matter are mentioned too late in the course, long after the student had internalized a very misleading, albeit initially helpful, image of atoms, electrons, etc. It is very difficult to undo the 'damage' and discard the evocative metaphors. I know, because I went through the process of learning and unlearning in successive cycles. I wish they had told me the truth up front.

And the truth is very simple: We don't know what protons, neutrons, electrons, etc. are – we only know how they manifest in our experiments. If we focus on experimental data, without building convenient pictures in our minds; if we keep an open mind and follow the data where it leads, we will have reliable knowledge of the world of atoms.

We will know what we know, how we know it and what are the limits of our knowledge.

Actually, these are the essential rules of critical thinking in any area of human endeavor – politicians, sociologists, and historians would be well advised to keep them in mind. I would like you to read the next section about the early steps of discovery with these thoughts in mind. If ever you are tempted to think of electrons as tiny planets orbiting the nucleus, I suggest you recall the difference between pulling off a band-aid slowly or with one quick jerk.

Crucial Experiments

After idle philosophical speculations by the ancient Greeks, the science of atomic physics finally got under way in the 17th century with Robert Boyle. He was among the first to insist on precise, repeatable, controlled experiments to determine the nature of reality, as he explained in his book *The Sceptical Chemist,* in 1661.

As mentioned in Thermodynamics:

In 1662 British physicist Robert Boyle, after developing the first efficient air pump, discovered an inverse relationship between the volume and the pressure of a gas: pV = const. This is called the Boyle-Mariotte Law today because French physicist Edmé Mariotte discovered the same law independently from Boyle 16 years later, adding the important observation that it is true only if the temperature of the gas is kept constant. Their findings were easy to explain if one assumed that a gas consists of small solid particles with lots of space between them. The volume would then be decreased by pressing the particles closer together. For this reason, Robert Boyle became one of the first 'modern' atomists.

In 1794 French chemist Joseph Louis Proust noticed a curious thing: elements seemed to combine into compounds according to specific proportions. When mixing elements that were known to form compounds (e.g. copper, carbon, and oxygen forming copper carbonate), one had to mix them together in these proportions if one wanted to use up all the ingredients. In Proust's words:

"I shall conclude by deducing from these experiments the principle I have established at the commencement of this memoir, viz. that iron like many other metals is subject to the law of nature which presides at every true combination, that is to say, that it unites with two constant proportions of oxygen. In this respect it does not differ from tin, mercury, and lead, and, in a word, almost every known combustible."

His results were confirmed by Swedish chemist Jacob Berzelius in 1804 and since then, the phenomenon has been called "**the law of definite proportions**".

In 1803 English chemist John Dalton announced a new law: it is possible for certain elements to combine together in more than one set of proportions, each producing a different compound with different properties (for example carbon monoxide and carbon dioxide). However, the law of definite proportions held for each separate compound. He called his new law "**the law of multiple proportions** ". He suggested that the two laws could be explained if we assumed that each element was made up of very small solid particles and revived the old name coined by Democritus: **atoms**. Compounds were made by these particles combining together in a specific configuration for each molecule of the compound. In 1808 he published a book titled *New System of Chemical Philosophy,* where he explained why assuming the existence of atoms was a good idea, in view of all the experimental data accumulated up to that time.

In 1827 Scottish botanist Robert Brown noticed that small pollen grains suspended in the water moved about in a zigzag fashion, randomly changing direction and speed, even though there was no current in the water. The same happened when any small enough pieces of matter were studied: they all moved around in the same way. This phenomenon is now called "Brownian motion" and it was Einstein who suggested a convincing explanation in 1905 (see later), based on the atomic theory of matter. Since no one could think of an alternative explanation, even the most hostile opponents of the theory admitted defeat.

In 1869 Russian chemist Dmitri Ivanovich Mendeleev suggested that the already known elements could be arranged in order of their atomic weight (as assumed at the time) in a table, such that elements of similar chemical properties would fall in the same column. This way, several gaps were left in the table and Mendeleev announced that three of these gaps represented elements not yet discovered. This is science at its best; a bold predict on is made based on a theory and, by 1885 all three elements were discovered.

Assumptions and theories

Once the existence of atoms was no longer doubted, further assumptions were made about their properties. After checking the logical consequences of these assumptions, a coherent theory was built to describe the nature and properties of the different atoms.

Chemists knew what elements a compound was made of if they obtained the compound by mixing elements together. Sometimes they could only guess, and they had to wait for some technique of breaking down the compound and analyze the component elements.

In 1800 British chemist William Nicholson, first succeeded, by passing an electric current through water, decomposing it into two known gases: oxygen and hydrogen. Examining the obtained gases he found that:

- The volume of hydrogen produced was twice the volume of oxygen
- The mass of hydrogen obtained was one-eighth the mass of the oxygen.

From these observed facts, the following conclusions were drawn:

- Each water molecule is composed of one oxygen and two hydrogen atoms
- The weight of an oxygen atom is 16 times the weight of a hydrogen atom.

These conclusions were based on further assumptions that the volume of hydrogen produced was twice the volume of oxygen because there were twice as many hydrogen molecules released than water molecules. This wasn't a proven fact as yet but seemed reasonable.

In 1811 Italian physicist Amedeo Avogadro made a suggestion that is now known as <u>Avogadro's hypothesis</u>, which states that a given volume of *any gas*, at the same temperature and pressure, contains the same number of molecules. This was a bold supposition that certainly explained Nicholson's two observations about water. If proven true, the hypothesis would be an enormous help in finding out how many atoms of each component element made up a molecule of a compound.

In 1828 Swedish chemist Jacob Berzelius published a table of atomic weights based on experimental data similar to the one I just described about the water molecule. This table was a definite improvement over Dalton's table, which incorrectly assumed, for example, that the oxygen atom was eight times as heavy as the hydrogen atom. Nevertheless, it still contained errors.

In 1858, Italian chemist Stanislao Cannizzaro convinced Europe's chemists of Avogadro's hypothesis, and that led to better and better atomic weight tables during the 1865 – 1905 period.

It is important to realize that scientists still had no idea what the mass and size of an actual atom were; they only knew the **relative weights**, based on the assumptions described above. Hydrogen was found to be the lightest of the elements, but it was oxygen whose **atomic weight** was arbitrarily selected by contemporary chemists as 16, that made the atomic weight of hydrogen a little bit more than one. Obviously, the weight of a million oxygen atoms will also be sixteen times the weight of a million hydrogen atoms.

To make chemical calculations easier, a new mass unit, the **mole** was defined as follows: **one mole of mass** of any substance, element or compound, equals the substance's molecular-weight number of grams. For example, one mole of oxygen molecules (O_2) has a mass of 32 grams, and one mole of hydrogen molecules (H_2)

has a mass of 2 grams. Since the oxygen molecule is 16 times as heavy as the hydrogen molecule, it is obvious that 1 mole of oxygen contains the same number of molecules as one mole of hydrogen. It is also true for all the elements: **One mole mass of any substance contains the same number of molecules.** This number is called **Avogadro's constant** and is denoted as N_A.

In 1865, Austrian scientist Johann Josef Loschmidt, using the kinetic gas theory, estimated the value of Avogadro's constant as $N_A = 6.023 \times 10^{23}$. In possession of the Avogadro number, Loschmidt also estimated the size and mass of the hydrogen atom as roughly one hundred-millionth of a centimeter, and a mass of 1.67×10^{-24} gram. These values were confirmed in 1913.

The scene was now set for Mendeleev and the Periodic Table he sprung on the scientific community in 1869. Having his theory dramatically confirmed by the discovery of the predicted elements in 1885, modern chemistry was born.

Kinetic Theory of Gases

Recall that in 1662 British physicist Robert Boyle discovered an inverse relationship between the volume and the pressure of a gas: pV = const.

This was one of the very first experiments to suggest that gases might be made of small particles (atoms) with space between them, so they can be compressed together. Even though other, even more convincing experimental results were still in the future ("the law of definite proportions" in 1794; "the law of multiple proportions" in 1803), one physicist was ready to accept the existence of atoms and draw impressive conclusions from that assumption:

In 1738, Swiss mathematician Daniel Bernoulli published a book, *Hydrodynamics*, in which he made the following assumptions:

- Gases are composed of a large number of atoms.
- There is empty space among these atoms.
- Atoms are in constant, fast, random motion.
- Atoms collide with each other all the time elastically.
- The pressure of a gas is due to atoms hitting the walls of the container.

Using these assumptions and Newton's laws of motion, he applied the probability techniques pioneered by Fermat and Pascal, and mathematically deduced Boyle's Law and even the Combined Gas Law (if we interpret temperature to be proportionate to the kinetic energy of the gas molecules). His assumption of atoms, however, was too wild for his contemporaries and his results were ignored for almost 100 years.

By 1858, when Italian chemist **Stanislao Cannizzaro** convinced Europe's chemists of Avogadro's hypothesis, the existence of atoms was accepted by almost everyone. Bernoulli's work was rediscovered and taken up by three of the most prominent theoretical physicists of the era: Clausius, Maxwell, and Boltzmann.

Clausius assumed that all air molecules travel roughly at the same speed, which had to be very fast if pressure was to be interpreted as the bombardment of the container walls by the air molecules. However, diffusion (for example the spread of a smell) was known to be relatively slow, so Clausius further assumed that the molecules collide with each other very frequently, constantly changing direction.

In 1859 Scottish physicist James Clerk Maxwell, having read Clausius's paper on the diffusion of gases, decided to treat the subject mathematically. A statistical method produced a distribution of the speed of molecules, instead of assuming they were all going at the same velocity. This distribution law described a Bell curve, the first of its kind in the history of physics. Maxwell's theory predicted a yet untested law, according to which the viscosity of gases should be independent of their pressure. A few years later this prediction was proven experimentally.

These first steps slowly built up our understanding of the internal structure of gases. It can be verified that the number of collisions a single molecule of gas receives in one second is about 10^{14}. There is a direct correspondence between the temperature of the gas and the average speed of the gas molecules: for example, at zero Celsius, the average speed of a hydrogen molecule is about 1.8 km/sec, while the average speed of an oxygen molecule is about 0.46 km/sec and that speed, in one mole of gas ($6.023 * 10^{23}$ molecules), in a small container, will guarantee a lot of collisions.

In the middle of the 19th Century, all this was pure speculation in some brilliant minds, awaiting confirmation by experimental data.

Ludwig Boltzmann (1844 - 1906)

Boltzmann is one of science history's tragic characters. He was brilliant and admired in his native Austria by his students, his colleagues, his superiors. Yet, he was chronically restless. He kept moving from one academic post to another, even returning to the same university three times during his 40-year career. In periods of deep depression, he often thought of killing himself and went as far as a failed suicide attempt n Leipzig in 1901.

In addition to bipolar disorder, he also had to deal with personal unhappiness: his revolutionary ideas were often viciously attacked by some of the most prominent physicists of his time; in particular by Ernst Mach and Wilhelm Ostwald. The basis of their animosity was the rejection of Boltzmann's central assumption about the existence of atoms, which was still controversial at the time.

The controversy over the reality of atoms, about which speculation started with Democritus as early as 460BC, kept minds busy for centuries. As experimental data and mathematical theories accumulated, this argument heated up at the end of the 19th century, with powerful proponents and opponents.

For example, John James Waterston, a Scottish physicist, attempted to publish a paper at the Royal Society in the 1840s on his kinetic (atomic) theory of heat, but his paper was rejected. It would have died in obscurity if Lord Rayleigh (English physicist, 1842-1919) had not found the rejected paper in the Society's archives in 1893. He appreciated the brilliant insight and convinced the Society to publish it, 50 years later. Rayleigh advised future generations of young physicists:

"The history of this paper suggests that highly speculative investigations, especially by an unknown author, are best brought before the world through some other channel than a scientific society, which naturally hesitates to admit into its printed record matter of uncertain value. "

Translation: if you are an unknown author don't count on the scientific establishment if you have revolutionary new ideas.

Contrary to some admiring myths about scientists, they are human like the rest of us. A famous quote by Max Planck is appropriate here: "A new scientific truth does not triumph by convincing its opponents and making them see the light, but rather because its opponents eventually die, and a new generation grows up that is familiar with it."

In 1868 Boltzmann started working on the kinetic theory of gases, trying to improve on Maxwell's theory of speed distribution of the molecules. He succeeded brilliantly and today it is called the "Maxwell-Boltzmann distribution". He spent the rest of his career perfecting the theory and by 1905, when he traveled to San Francisco to lecture on the subject, he had it all completed mathematically. The result is the incredible breakthrough of interpreting **entropy** in a statistical way, as the measure of disorder or, more precisely, the logarithm of the probability of disorder.

$$S = k * \ln w$$

This is the Boltzmann's Equation, which means that the entropy of a system (S) is proportional to (the natural logarithm of) the probability (w) of the disorder in the energy distribution of the gas molecules. The proportionality factor (k) is called the "Boltzmann Constant" with a magnitude of 1.3807×10^{-23}.

This equation is the crowning achievement of Boltzmann's contribution to physics and it is quite appropriately carved on Boltzmann's tombstone in Vienna.

The philosophical interpretation of the Boltzmann Equation is well known. We live in a universe that is in a low-entropy state, meaning a very uneven distribution of matter and energy, concentrated in galaxies and stars that constantly radiate energy into the vast empty space in between. Since entropy cannot decrease in a closed system (and the universe is closed by definition), the energy of the stars will, in time, dissipate and reach maximum disorder in an equilibrium of maximum entropy. Over-dramatic physicists call this final state the "heat death of the universe" in which life can no longer exist.

Boltzmann died prematurely: in 1906, during a summer holiday in Trieste, he committed suicide.

Brownian motion – the triumph of atomism

It was Albert Einstein, with his paper on Brownian motion in 1905, who finally settled the debate on the atomic hypothesis.

In the *Annalen der Physik*, 17, p. 549, 1905, he wrote:

"If the movement discussed here can actually be observed … then classical thermodynamics can no longer be looked upon as applicable with precision to bodies even of dimensions distinguishable in the microscope: an exact determination of actual atomic dimensions is then possible. On the other hand, had the prediction of this movement proved to be incorrect, a weighty argument would be provided against the molecular-kinetic conception of heat."

What an admission! As recently as 1905, even Albert Einstein wasn't absolutely certain that the kinetic (molecular) theory of heat would prove to be correct! However, in a later paper (Annalen der Physik (4) 19, 1906 pp. 371-381) he triumphantly reports:

"Siendentoph (of Jena) informed me that he and other physicists ... had been convinced by direct observation that the so-called Brownian motion is caused by the irregular thermal movement of the molecules of the liquid....Not only the qualitative properties of the Brownian motion but also the order of magnitude of the paths described by the particles correspond completely with the results of the theory."

In 1908, the French physicist Jean Baptist Perrin proved Einstein's theory by meticulous experimental observation. In 1909, Friedrich Wilhelm Ostwald, Russian-German physical chemist (Nobel prize recipient in 1909), the last illustrious doubter of atomic theory, capitulated: he admitted that atoms did indeed exist!

The Elements (atoms)

Students of physics, as a rule, don't use much chemistry. Some physicists even go to the length of declaring chemistry a branch of applied physics, completely explained by Quantum Mechanics. The absolute minimum physics students need is an understanding of, and familiarity with, the Mendeleev Periodic Table. This table was based upon experimental facts accumulated by chemists over the centuries. Knowing the basic elements and their physical/chemical properties is basic science that everyone, not only students of physics, should acquire.

So what are these elements we (and the universe) are made of? Some of them we know quite well from everyday experience: iron, copper, gold, silver, aluminum, carbon, oxygen, neon, titanium, zinc, nickel, chromium, chlorine, mercury, platinum, calcium, lead. Seventeen elements out of 118 known to exist today. Mostly metals with a few non-metals and gases thrown into the lot. Most are solids at room temperature, some of them are liquid and some are gaseous.

Most of these elements can be found in nature, most of them were known to our forefathers in antiquity. The others were discovered over the centuries by busy alchemists, chemists, physicists, engineers, and tinkerers of all kinds. In modern industrial technology, most of the known elements are used one way or another to manufacture cars, houses, clothes, appliances, books, Christmas ornaments, Barbie dolls, drugs, and food stuff. The list is endless.

Most of us don't know what elements are used for what, who discovered them, where and how they are obtained, what beneficial and harmful effects they can have on our health and our environment. However, physicists need to know that they are all made up of different atoms: different in size, in mass, in internal composition, in the ability to conduct heat and electricity, and to react with other kinds of atoms.

The magnificent book: "Elements" by Theodore Grey defines the Periodic Table of the chemical elements as a "universal catalog of everything that you can drop on your foot". An apt definition indeed, rephrasing the well-known and quite an ancient definition of: "everything the world is made of" - including our bodies. They are inside our bodies and we wouldn't exist without containing any of the following (% by mass):

1. Oxygen (65%)
2. Carbon (18%)
3. Hydrogen (10%)
4. Nitrogen (3%)
5. Calcium (1.5%)
6. Phosphorus (1.0%)
7. Potassium (0.35%)
8. Sulfur (0.25%)
9. Sodium (0.15%)
10. Magnesium (0.05%)
11. Copper, Zinc, Selenium, Molybdenum, Fluorine, Chlorine, Iodine, Manganese, Cobalt, Iron (0.70%)

12.   Lithium, Strontium, Silicon, Lead, Vanadium, Arsenic, Bromine (trace amounts)

In addition to the fascinating subject of what we and the world are made of, it is useful to know what we are using in our daily lives and how it may affect the rest of the world. Elements are identified by a number of defining characteristics. These are:

Examples of Physical Properties
- Color
- Luster
- Hardness
- Odor
- Luminescence
- Conductivity
- Solubility
- Malleability
- Ductility
- Density
- Viscosity
- Compressibility
- Freezing point
- Boiling point
- Melting point
- Crystalline structure
- Allotropic

Examples of Chemical Properties
- Flammability
- Toxicity
- Radioactivity
- Heat of combustion
- Reactivity with water
- Reactivity with acids
- Oxidation
- Corroding

The 118 known elements are not all totally unique. Many of them are so similar to each other that chemists had a hard time finding the differences. Elements with similar physical and chemical properties are often grouped together into families.
These families are the following:

Hydrogen (H, 1) – Sunshine, Hindenburg, Fuel cells

Alcali Metals
- low melting and boiling points compared to most other metals
- very soft and can be cut easily with a knife
- low densities (lithium, sodium, and potassium will float on water)
- react quickly with water

Examples :
- Lithium (Li, 3) – Batteries, Mood-swing pills
- Sodium (Na, 11) – Salt, light bulbs, headlights
- Potassium (K, 19) – Fertilizer, Medical pills

Alkaline Earth Metals
- shiny.
- silvery-white.
- somewhat reactive at standard temperature and pressure.
- low densities.
- low melting points.
- low boiling points.

Examples:
- Magnesium (Mg, 12) – Photo flash, campfire starter
- Calcium (Ca, 20) – Chalk, bone, gypsum

Transition metals
- form colored compounds.
- good conductors of heat and electricity.
- can be hammered or bent into shape easily.
- less reactive than alkali metals such as sodium.

- high melting points

Examples:
- Titanium (Ti, 22) – golf club, razor, tools.
- Vanadium (V, 23) – tools, drill bits, etc
- Chromium (Cr, 24) – cutlery, pots, car parts, bumper coating
- Tungsten (W, 74) – welding electrodes, glass drill, light bulbs
- Iron (Fe, 26) – everything, everywhere
- Cobalt (Co, 27) – drills, cobalt glass
- Nickel (Ni, 28) – coins, bumper plating, stainless steel
- Platinum (Pt, 78) – Spark plugs, jewelry, lab equipment
- Copper (Cu, 29) – wires, plumbing, coin, jewelry, crafts
- Silver (Ag, 47) – Coins, jewelry, mirrors, electronics
- Gold (Au, 79) – Coins, jewelry, electronics
- Zinc (Zn, 30) – Coins, galvanization, batteries
- Mercury (Hg, 80) – thermometer, dentistry, light bulbs

Metalloids
- Solid
- Metallic luster
- Brittle
- Semi-conductive.
- Average transmission of heat
- Readily form glasses
- Form alloys with metals
- Several metallic and non-metallic allotropic forms
- Some metalloids contract on melting
- Reacts with the halogens to form compound

Example:
- Silicon (Si, 14) – glass, sand, semiconductors

Halogens
- non-metals
- low melting and boiling points

- brittle when solid
- poor conductors of heat and electricity
- colored vapors

Examples:
- Fluorine (F, 9) – Teflon, medical use
- Chlorine (Cl, 17) – Salt, bleach, disinfectant
- Iodine (I, 53) – medicine, disinfectant

Noble gases
- non-metals
- very unreactive gases
- colorless

Examples:
- Helium (He,2) – ballons, lasers
- Neon (Ne, 10) – advertising, lasers

These are the main groups as we understand them today, but they were not identified as such before Mendeleev developed his periodic table. However, similarities among chemical elements had been known by chemists (and even alchemists) for centuries.

Our technological environment contains most of the 118 known elements and most of us wouldn't know what they are. An example of an element unknown to the average citizen is Niobium (Nb, 41). Take the ever so popular cell phones most of our children couldn't live without today (at least they think so). They contain the element Niobium, or Tantalum - dense, heat-resistant, non-corrosive metals that hold an electric charge well – essential in cell phones (Niobium, discovered in 1801 by Charles Hachett, is also used in body-piercing and decorative coins). Where do these metals come from? The largest supplier (60%), until recently, was the Democratic Republic of Congo in Africa. When cell phones became the rage, prices shot up, Congolese peasants abandoned their fields to dig up the thick mud (called "coltan") in creeks that contained the metal to sell to the white people. Unfortunately, the abandoned fields did not produce the crops they needed to eat, so they began hunting for gorillas for meat, practically exterminating the already endangered population. So, when we use our cell phones or indulge in body piercing, we are, in a small way, responsible for hastening the extinction of a magnificent, irreplaceable species on our planet.

Niobium has another, even deadlier application. Because it is highly resistant to corrosion, even at extremely high temperatures, it is used in rocket nozzles in missiles. Amusingly (not for him), Theodore Gray, the author of "Elements", and an avid 'element-collector', was visited by the FBI to confiscate his niobium sample that he had acquired from eBay. Apparently, it was a quite up-to-date missile part.

.   .

So, what are these elements doing in a book on physics? The answer is simple and obvious: these elements are all atoms: unique, different. They obey very strict rules that took mankind centuries to discover. Before we go into details of these rules, we need to learn what these atoms are, how they are related to each other, how they can be organized in a logical system that shows the underlying pattern of nature's building blocks. This quest will bring us to our first genius: Dmitri Ivanovich Mendeleev (1834-1907), a Russian chemist, the father of the "Periodic Table" of chemical elements.

Dmitri Ivanovich Mendeleev (1834-1907)

Mendeleev was born in the western Siberian town Tobolsk. His father was the headmaster of the local gymnasium (high school) who became blind in the year Mendeleev, the youngest of fourteen children, was born. His mother supported the family by managing a local glass factory until it burned down in 1848, one year after Dimitri's father died. Her older children grown up by then, Maria Mendeleev decided to take her youngest son and his still young sister to Moscow to give him the opportunity of a university education. This was a 2000 km journey, undertaken in a horse-drawn wagon. It took them whole summer to get there.

Dimitri was refused admittance to the University of Moscow because of the existing quota system that did not include Siberia, so his mother loaded her children up on the wagon and traveled with them another 640 km to St. Petersburg university where she got the same answer as she had received at the University of Moscow. Luckily, she met an old classmate of her husband who managed to enroll him in the Central Pedagogical Institute – a training college for high-school teachers. Mendeleev also received a small government scholarship to cover his expenses while studying. He needed that support because his mother died within a year after they arrived in St. Petersburg, soon followed by his sister. They both had tuberculosis.

Mendeleev was now alone in St. Petersburg, without family or friends. He buried himself in science and soon became the top student. He graduated in 1855 as the best student of the year and was sent to Simferopol in the Crimea for his first teaching position. He returned to St. Petersburg in 1856 and, due to his academic excellence and already published scientific papers, he was appointed as an unpaid lecturer at the University of St. Petersburg.

In 1859 he received a government grant to study abroad for two years. He first went to Paris where he studied under Henry Regnault, then proceeded to Heidelberg, attending the classes of Kirchoff and working in the laboratory of Robert Bunsen – the inventor of spectroscopy. Because of Mendeleev's temperament (volatile to say the least) he and Bunsen didn't get along, so Mendeleev set up his own modest laboratory in his lodging where he started original research on the solubility of alcohol in water – the subject of his doctoral thesis.

The most important event during his study abroad happened in 1860 when he attended the first-ever international chemistry conference in Karlsruhe, Germany. The congress was called to resolve the confusion regarding atomic weights. There had to be a uniform standard, so the international community of chemists could communicate with each other, using the same measurement units. For example, everyone knew that water was composed of eight parts of oxygen and one part of hydrogen *by weight*. So many assumed that the atomic weight (relative weight to hydrogen) of oxygen was 8. They referred to it as "equivalent weight". Many used the two concepts interchangeably, and often confused the concept of "atomic weight' with "molecular weight".

After much wrangling, the problem was resolved by Stanislao Cannizzaro, an Italian chemist, who proposed to accept Amedeo Avogadro's 1811 hypothesis, according to which under the same pressure and temperature each gas contains the same number of molecules. Avogadro also assumed that both oxygen and hydrogen gas were composed of two-atom molecules, instead of single atoms as many chemists imagined.

This hypothesis was based on the experimental fact that two volumes of hydrogen molecules plus one volume of oxygen molecules produced exactly two volumes of water molecules *with nothing left over*. If we assume, with Avogadro, that both oxygen and hydrogen gases were made up of 2-atom molecules, then this experimental fact can be stated in shorthand notation as:

$$2L \text{ of } H_2 + 1L \text{ of } O_2 = 2L \text{ of } H_2O$$

Now, if we also assume, as Avogadro did, that there are the same number of oxygen, hydrogen, and water molecules in the same volume, say 10 for argument's sake, then the above volume equation can be transcribed into the molecule-number equation as follows:

$$20 \ H_2 + 10 \ O_2 = 20 \ H_2O$$

If we count the atoms on both sides of this equation, we will see that there are 40 hydrogen atoms and 20 oxygen atoms on both sides. This result justifies both of Avogadro's assumptions.

Once we accept this hypothesis, arriving at the exact relative atomic weight of the elements would be easy. For example, for oxygen, all they had to do was fill two identical volume containers (under the same temperature and pressure) one with hydrogen, and the other with oxygen, and weigh them. Doing this they found that the oxygen weighed sixteen times as much as hydrogen, therefore the relative atomic weight of oxygen had to be 16 (as opposed to 8 as many chemists had assumed).

The chemists at Karlsruhe accepted Cannizzaro's suggestion because it made perfect sense. This way they had a uniform method for establishing atomic weights and use the same notation. The chemical notation for the water molecule became H_2O, eliminating the often used HO notation.

Mendeleev was one of the enthusiastic supporters of this convention and when he returned to Russia in 1861 he made sure that all his students were taught the new system of establishing atomic weights.

In the same year, he obtained a contract to write his own textbook on organic chemistry and finished the 502 pages manuscript in two months. The book was a success and he obtained the prestigious *Domidov Prize* for his work from the Russian Academy of Science. The prize money enabled him to propose marriage to his first wife Feozova Nitkitichna Leschev, a shy and admiring friend of his sister, and his chaotic bachelor home life obtained some order in a bigger apartment, better furnished and organized.

He continued to immerse himself in research, teaching, and writing and, as a result, in 1867 he became the professor of inorganic or general chemistry at the University of St. Petersburg. Mendeleev had it made career-wise, but he never settled down to quietly enjoy his accomplishments. Far from it.

In 1868 he published the first volume of his next textbook: "Principles of Chemistry", and that's where the saga of Mendeleev's periodic table began.

There were 63 known elements at that time and Mendeleev covered only 8 in the first volume of his new textbook, which was organized by the existing practice of chemical experiments and 'natural history' of the elements.

Mendeleev was in trouble. He had 55 elements left to deal with for the second volume and he knew that he would never finish it in time to meet his contractual obligation to the publisher if he followed the same method. He had to find a way to organize the elements in a more systematic way, so he started thinking about listing them in the order of their atomic weights. Isn`t it ironic? One of the greatest discoveries in chemistry started out by a need to somehow finish publishing a first-year chemistry textbook in time.

Mendeleev wasn't the first to think of organizing elements in some systematic way, not even the first who did this by atomic weights. In 1820 Johann Wolfgang Doebereiner found similarities between some elements in groups of three and called them "triads". He found that if he added the atomic weight of the first and the third, divided the result by two, the number obtained was very close to the atomic weight of the second. Thus, for the first time, a relationship between properties and atomic weights was suggested.

In 1862 Alexandre Emile Beguyer, a French geologist, arranged the elements in a spiraling column and found that Doebereiner's triads fell into the same vertical column.

In 1864 an English chemist John Alexander Reina Newlands divided the elements (by atomic weight) into columns of seven elements in each. He also found a pattern of similarities among elements in the same column.

And, finally, in 1870 a German chemist Julius Lothar Meyer arranged the elements, not by their atomic number, but by their properties – in particular, their specific gravity (the element's weight compared to the same volume of water) and produced a chart comparing atomic volume with atomic weight. He found a specific periodicity in his chart. He published his chart in 1870, one year after Mendeleev published his "first try" of his periodic table. Had he been a bit faster, now it could be called the "Meyer Periodic Table". Luck, as in all endeavors in life, plays an important role in science as well.

Mendeleev tried to organize the elements by their most important property: their 'valence' or combining power. It had been known that every element's atom had a maximum number of other atoms it can combine with into a compound molecule. As a first try, he listed the elements in order of their atomic weight and wrote down their valence next to each. He noticed that the valence numbers go up and down as the atomic weight increases.

The valence numbers of the first 17 elements listed by atomic weight were: 1,1,2,3,4,3,2,1,1,2,3,4,3,2,1,1,2. He definitely had something here. The first full cycle (if we don't include Hydrogen) had seven elements, the next also has seven.

After some trial and error, he created a table of elements with seven rows and seven columns, atomic weight increasing across a row and then continuing in the same increasing order from row to row. The first column had only one element in it: hydrogen. The second column had 7 elements: Lithium, Beryllium, Boron, Carbon, Nitrogen Oxygen and Fluorine. The third column had Sodium, Magnesium, Aluminium, Silicon, Phosphorus, Sulfur, and Chlorine.

He had to leave gaps in these columns, so the fourth, fifth, sixth, and seventh columns, with a lot more elements, could fit into a table structure. When he was done, each row of the table contained elements with similar chemical properties and could be considered belonging to the same 'family' (in our currently used version of the table, Mendeleev's rows became columns for easier presentation).

This is what makes Mendeleev the genius of the Periodic Table. He declared, with unbelievable confidence, that the gaps he had to leave in his table represented elements *not yet discovered*. In addition, to explain why some elements did not fit into his table, he declared, that the measured atomic weights of these elements were `wrong` and needed to be corrected!

Such chutzpa! If correct, it would make him world-famous. If wrong, it would bury him in shame and obscurity. He was not wrong. His predictions were proven correct, one after the other. For example, Beryllium was supposed to have an atomic weight of about 14. This wouldn't have fit into Mendeleev's table, so he said: it must be about 9 instead. After carefully repeating the measurements, it was found to be 9.013. He also predicted the atomic weights of indium and uranium were wrong and gave the values they should have to fit into his table. Again, he was right. He not only predicted the existence of the missing elements but described, in minute detail, their properties.

The first of the three predicted elements were found by French chemist Paule Emile Lecoq de Boisbaudran in 1874 that he named `gallium` after his country`s Latin name.

The second predicted element was found by Lars Fredrick Nilson of Sweden in 1882 and he named it `scandium' after his native Scandinavia.

In 1886, finally, the third missing element was found by Clemens Alexander Winkler, a German chemist and he named it `germanium` after Germany.

Mendeleev was now world-famous and his periodic table was accepted as a world-class discovery. He received the `Davy Medal` from the Royal Society of London in 1882, and he was showered with other medals and international recognition for his accomplishment.

His personal life was equally tumultuous – two episodes that no biographer fails to mention are so typical of the man that I can't resist repeating them here.

In 1876 at the age of 42, he fell in love with a seventeen-year-old art student. His own marriage was over by then, he and Feozova lived separated, if not yet divorced, He pursued Anna Ivanovna Popova relentlessly and, after he obtained a divorce from his wife in 1882 he bribed the local prelate with 10,000 rubles to marry them, even though Orthodox church law demanded that he wait seven years to marry again. When this transgression was brought to the attention to the czar, Alexander III, he has reportedly said: "I admit Mendeleev has two wives, but I have only one Mendeleev".

The other episode took place on the 7th of August 1887, near Mendeleev's summer home. He got permission, during a solar eclipse, to ascend in a balloon, piloted by military officer A.M. Kovanko. However, when the time came, the balloon failed to lift from the ground due to the heavy scientific instruments Mendeleev loaded up in order to make measurements of the sun's corona. Time was critical because the eclipse would last only a few minutes, so Mendeleev, without prior ballooning experience ordered his pilot out of the gondola and ascended to 500 meters, and made the observations alone. He had no problem controlling the balloon and managed to land safely on a field, a long distance from his original position, terrifying the peasants at work on their fields.

Mendeleev spent the rest of his life in scientific research, teaching, writing, and fighting the czarist bureaucracy, primarily on behalf of his students and the freedom and fairness of academic education.

While Mendeleev was busy with his elements and his table, scientists in the same period were busy with experiments, preparing for the major discoveries of the next century. Mendeleev knew about many of these experiments and used many in his own research. Some of them he greeted with skepticism, while a few others he refused to believe to be real.

Let's review the most important of these before we get into the next major saga of Marie Curie and Radioactivity.

Spectroscopy **(Kirchoff, Bunsen, 1860)**

It was one of the simplest inventions in the history of physics, yet it revolutionized almost all the sciences, from chemistry to astronomy.

Two names are connected to this discovery: Gustav Kirchoff (1824 – 1887), German scientist, and Robert Bunsen (1811 – 1899) German chemist. Both were professors at Heidelberg University. It was Kirchoff's idea: to help Bunsen study photochemistry (the ability of chemical reactions to absorb or emit light) he suggested to Bunsen that he should use a prism for the purpose.

It had been known since Newton's time that sunlight directed through a small slit in a surface and allowed to hit a prism produced a rainbow spectrum on a screen placed behind it. What Bunsen and Kirchoff found was that replacing sunlight with a light source of heated hydrogen gas did not produce a continuous spectrum, but several multicolored lines on the screen that was black everywhere else, except for these very unique lines.

They soon discovered that every element, when heated, produced a very distinct pattern of lights. These lines were called "emission lines" for obvious reasons.

A small variation of this experimental setup produced a different kind of spectrum that we today call "Absorption Spectrum". When we replace the light source with a light bulb, but interpose a cool cloud of hydrogen gas between the light bulb and the slit, on the screen now we will have a continuous rainbow spectrum, with dark lines overlayed in the exact same position of the emission lines of hydrogen.

With this method chemists very soon found many new elements by studying the spectral lines they obtained from their sample minerals. It also led to the discovery of Helium, years before the element was found on Earth, by studying the spectral lines of the sun itself (the reason it was named 'Helium').

The distribution of these spectral lines was of some interest and it was a Swiss mathematician and physicist Johann Balmer (1825-1898) who came up with a mathematical formula that described this distribution for the hydrogen gas. In 1885 he published his empirically obtained formula, now called the Balmer series for the wavelength distribution, without a physical explanation attached to it. The very fact that the distribution could be described by a mathematical formula proved that it was not random. It had to have some underlying principle in physics. This explanation had to wait for Niels Bohr, 28 years later.

X-Rays (Roentgen, 1895)

Physicists play with everything, all the time. Physics textbooks usually explain science in neatly summarized categories, like Mechanics, Electrodynamics, Thermodynamics, etc, but in real life, all of it happened at the same time. So, while Mendeleev was playing with elements, Kirchoff and Bunsen with spectroscopy, Wien with radiation, a German inventor Heinrich Geissler (1814 – 1879) played with vacuum tubes that became known as Geissler tubes, after their inventor. He succeeded in 1855 in producing the first vacuum tube that eliminated most of the air from the vessel (usually a glass tube sealed at both ends), allowing physicists to experiment with it.

Many physicists had tried to send an electric current through gases inside these tubes. They sealed two metal electrodes at the two ends inside the tube and connected them to two terminals of a battery. Even Faraday had tried it, but the tube he had was not good enough to produce new discoveries, although he did observe some fluorescence. Once Geissler tubes became available, physicists could observe weird patterns of light inside the tube as the current started and, when they included various minerals and gases inside the tube, they observed that the tube glowed in bright colors. Today we call them neon lights after the neon gas that was used in some of these tubes. They did not know what conducted electricity between the two electrodes inside the tube (they called them cathode and anode) so they called it "Cathode Rays".

This was at first mere curiosity, but then it took off in a big way when William Roentgen (1845-1923) German Physicist took an interest in the phenomenon. In 1895, trying to observe the luminescence appearing in his cathode ray tube, he darkened the room, covered most of his vacuum tube with black material, and turned on the power. Unexpectedly he noticed a flash of light at the other end of the room. That light came from a sheet of paper that was covered with a barium compound. When he turned off the power, the light disappeared. Turning it back again he noticed the same thing. He experimented with different distances and obstacles between his tube and the paper and it worked every time, even when he took the paper to the other room. He knew that it could not be caused by some of the Cathode Rays escaping from the tube because putting a magnet between the tube and the paper made no difference and, by that time, it was known that magnetic fields deflect Cathode Rays.

Roentgen didn't know what kind of radiation he discovered, so he called it "X-Ray" radiation. He knew that he was on to something new and important and spent the next seven weeks experimenting and documenting the properties of this new radiation. When asked later what he thought of it at the time he replied: "I didn't think, I experimented". Physicists always worry about others beating them to their discovery, so it is a race to produce a scientific paper with enough data before someone else does. In 1896, in a public lecture, he produced a photograph of someone's hand, showing all the bones inside.

This discovery gave a major boost to science, medicine, and technology, still widely used today.

Becquerel rays (1896)

Antoine Becquerel (1852-1908) French physicist stumbled upon radioactivity the same way Roentgen discovered X-rays – by accident. He was studying the fluorescence of different materials when he learned about Roentgen's discovery.

Luminescent materials had been known to exist for many centuries, even the Chinese wrote about them in 1500BC. Glow-in-the-dark materials (photo-luminescent) were studied seriously from the 17th century on and, after careful observation, two different kinds were identified. One type, when exposed to sunlight for a while, continues to glow in the dark for varying lengths of time. These materials are called "Phosphorescent". The other gives an immediate flash as soon as it is exposed to strong light but stops glowing instantaneously when the light source is removed. These materials are called "Fluorescent".

Becquerel studied both and, having heard about X-Ray's fluorescent effect on some materials, he wondered if fluorescent materials could be a source of X-Rays. So, in 1896, he set out to find the answer. His experiment was very simple: He wrapped a photographic plate in black material, placed a fluorescent crystal (containing uranium atoms) on top of it, and exposed both to sunlight. He reasoned that if fluorescence (to be generated by the sunlight) contained X-Ray elements, then the X-Ray's penetrating power would cloud the photographic plate.

True enough, it happened, and Becquerel was convinced that he was right. However, during cloudy days, when there could not be any fluorescence, the photographic plate also clouded, even though his experiment had been inside a dark drawer in his desk for days. It appeared that the fluorescent material he used kept emitting X-Ray-like radiation continuously, without being given any energy from the sunlight or any other source. This was amazing because it seemed to contradict the conservation of energy principle that had been firmly established by this time.

Further experiments proved that the new kind of radiation he discovered could be deflected by magnetic fields, so it wasn't X-Ray after all, but what was it?

The electron (1897)

Joseph John Thomson (1856-1940) English Physicist was fascinated by cathode rays. In particular, he wanted to know why these rays reacted to magnetic fields, but not to electric fields. Nobody knew anything about the nature of these rays, even though speculations were rampant. The most common assumption was that cathode rays were electric current through the aether, not carried by actual negatively charged particles. This assumption was reinforced by the fact that nobody had been able to deflect the rays with an electric field.

J.J. Thomson (as he was routinely called) wanted to find out. He suspected that maybe previous experimenters did not use a strong enough electric field, or the vacuum tubes were not evacuated enough. His experiment was ingenious. He subjected the rays to *both* electric and magnetic fields *at the same time* and measured the deflection that was noticeable due to the magnetic field, but not the same amount as had been predicted by the Lorentz force field equation. The difference was tiny, but there was a difference and the only reason for it could be the electric field because when he turned it off, the difference disappeared. He had confirmation: a cathode ray was a current carried by electrically charged particles.

What kind of particles? Where did they come from? What was their mass, their speed, their charge? Answering these questions was his next task.

From the amount of deflection, J.J. could only calculate the charge over mass ratio and, since he did not know either the mass or the charge of the particles, it was no help. However, he made a few assumptions: he knew the charge of the positive Hydrogen ion and assumed that Hydrogen, being the smallest atom, its ion must have the smallest charge possible. So, if the cathode ray particles were carrying the same charge, then he could estimate their masses from the e/m ratio he already had.

The calculation showed him that it was less than one-thousandth the mass of the atom. The only place these particles could have come from was inside the atom itself. This proved for the *first time ever*, that atoms were not the smallest, and indivisible, particles in the universe, that atoms had an internal structure and contained very tiny sub-particles. J.J. Thomson announced his findings at the Royal Society on April 29, 1897. Particle physics was born with the cathode ray particles, later called 'electron'.

The electron's charge was confirmed independently in 1909 by Robert Andrews Millikan (1868-1953), an American Physicist, with his famous experiment in which he sprayed oil droplets in between two plates of a charged condenser. He had to focus on one of the hundreds of very tiny droplets, looking through a microscope aimed at the gap between the plates, and from the speed of the oil droplet, he could calculate the charge of a single electron. Of course, the charge of the ionized droplet was due to more than one electron but, as he repeated the experiment hundreds of times, the accumulated data showed that the measured and calculated charges were always exact multiples of one, and only one, value and that had to be the smallest charge: that of the electron. I remember this experiment very well from my graduate student days because it was our nightmare experiment: the slightest noise, like someone coughing, or closing the lab door caused a vibration that scrambled the oil droplets in the microscope's lens and we had to start all over again.

Radioactivity (1898)

The next major discovery in particle physics was made by Marie Curie (born Marya Sklodowska) Polish-French chemist and her contributions are so important that it requires a longer exposition.

Marie Sklodowska Curie (1867-1934)

She was 17 when she finished high school in Russian-dominated Poland, first in her class. By that time, she spoke fluent Polish, Russian, German, French, and English. Women were barred from entering university and she had to look for some occupation to earn a living to help her widowed father and, most of all, support her older sister who had gone to Paris to study at the Sorbonne to become a doctor.

The two sisters made a pact: Marie would support Bronya during her education and, when she was established with a degree and an income in Paris, Marie would join her to start her own education.

It took nearly eight years for that dream to come true. The only way she could earn a living in Poland was to hire herself out as a governess, far away from her home in Warsaw, and live with rich families, teaching their spoiled children and putting up with the condescending humiliation by the parents.

In her spare time, she taught herself science and mathematics, from borrowed books, and waited for her turn. She was 24 when she made it to Paris, at the age when most great scientists have already finished their education, and many have already accomplished their greatest discoveries.

Her sister, just starting out as a doctor, married and with a child, had very limited means and could not help much. Marie lived in Paris in poverty so deep that we find it hard to imagine today. She lived in an unheated attic where the water froze during the night in her pitcher. She regularly fainted with hunger in class and could not mail a letter home for weeks because she could not afford to buy a stamp.

Yet, she obtained her Master's degree, at the top of her class, in two years - Physics in 1893, Mathematics in 1894.

In 1895 she married Pierre Curie, an already accomplished physicist who, with his brother Jacque, discovered piezoelectricity (the property of some crystal to create an electric potential under pressure) and was well known internationally for his other discoveries and inventions. They were both very poor, living on Pierre's teaching salary which barely covered their minimum expenses. The wedding was not performed in a church since both were free thinkers. She was given a wedding dress as a present by her sister's mother-in-law and Marie's only request was: "If you are kind enough to give me one, please let it be practical and dark so that I can put it on afterward to go to the laboratory". She had only one dress of her own. They spent their honeymoon on their used bicycles, biking through the country, sleeping in cheap quaint inns on their way. Their first child, Irene (who would become a Nobel prized Physicist herself), was born in 1897.

To finish her education, Marie needed to get her doctorate. For that, she had to select a topic where she could do new and original research. She was fascinated by Becquerel's discovery of spontaneous radiation produced by uranium compounds. Nobody knew how it was possible, where the energy came from. She decided to research this new and exciting subject.

It was easy for Einstein and other theoretical Physicists: they did not need a laboratory. All they needed was uninterrupted time, so they could think and calculate. Marie wasn't that lucky. She needed a place where she could do her experiments.

She was given access to a storeroom in her husband's school, without proper heating (the temperature went down to 6 degrees Celsius during winter). There she used her husband's inventions to study the nature of radiation. She soon discovered that the amount of radiation that she could measure precisely with the help of a piezoelectric crystal was proportional to the amount of uranium contained in the sample compounds she studied. The amount of radiation was also independent of environmental factors and the chemical composition of her samples. She soon came to the conclusion that the phenomenon had to be an atomic property.

The next logical question was: could this radiation be a unique property of uranium only, or did other elements or compounds possessed it? She decided to examine all known chemical substances and soon found another candidate: the element thorium also emitted similar spontaneous radiation. She discovered a new Physical phenomenon and she named it *radioactivity*, and the elements possessing this ability she named *radio elements*.

The experiments continued. The first hint of a major breakthrough came when Marie realized that the radiation she measured emanating from the uranium compound Pitchblende was four times stronger than the pure oxide of uranium that she could extract from it. This suggested that there had to be another new substance, besides uranium, contained in the compound that was also radioactive. Since conventional chemical analysis could not detect it, the conclusion was: it had to be a very small percentage of the compound, therefore a much more powerful radioactive substance than uranium itself.

On April 12, 1898, in the Proceedings of the French Academy of Science, it was announced that Marie Sklodowska Curie discovered in the pitchblende ore of uranium the probable presence of a *new element* possessing radioactive properties.

Pierre got so excited by his young wife's discovery that he abandoned his own research and joined her in pursuing this revolutionary discovery. They estimated that the ore should contain this new element in less than one percent. Had they known that the concentration of this new element was only a one-millionth part of the ore, they might have been concerned about their chances for success.

After they separated all the elements in pitchblende ore, they found that there had to be two different substances, with different strengths of radioactivity instead of one they had assumed before. By July 1898 they were ready to announce one of these to the Academy:

"We believe the substance we have extracted from the pitchblende contains a metal not yet observed, related to bismuth by its analytical properties. If the existence of this new metal is confirmed, we propose to call it polonium, from the name of the original country of one of us".

On October 17, 1898, they made a similar announcement:

"…The various reasons we have just announce led us to believe that the new radioactive substance contains a new element to which we propose to give the name radium".

Now all that remained was to isolate these new elements and measure their chemical properties, most of all their atomic weights. Without that, chemists and physicists all over the world would remain curious but skeptical about these announcements.

It took the two of them four years of backbreaking hard labor, processing eight tons of pitchblende ore residue, to obtain one gram of radium. They had no help, no laboratory, and no funding. The pitchblende ore was very expensive, and they could not afford it from their own savings. However, Marie reasoned: we don't need the uranium, maybe we can get the industrial waste after the uranium has been extracted from the ore? They contacted a mine in Austria and got a positive response: if they paid for the transportation, they could have the waste product for free. The first ton of it was soon delivered to their new 'laboratory' – an unheated shed with a leaking roof and dirt floor.

Mary wrote: "I came to treat as much as twenty kilograms of matter at a time, which had the effect of filling the shed with great jars full of precipitates and liquids. It was killing work to carry the receivers, to pour off the liquids and to stir, for hours at a stretch, the boiling matter in a smelting basin".

They continued their research and experiments and published more and more papers on the properties of these new elements, but the isolation of pure radium was not accomplished until 1902 when Marie finally extracted a tenth of a gram of pure radium and was finally able to measure its atomic weight: 225.

Radium officially entered Mendeleev's table. It was a white powder that glowed beautifully in the dark. Chemists and physicists started to study it in a serious way. In 1903 Ramsey and Soddy (English scientists) found radium spontaneously releasing a small amount of Helium gas. *This was the first indication of atoms spontaneously transforming into other elements*. The alchemist would have danced with joy! Other properties quickly emerged. Its radiation was two million times stronger than uranium's. Its penetrating power was phenomenal: only a thick block of lead could stop the rays emanating from radium. They analyzed the rays and found that they contained three different kinds: alpha, beta, and gamma as they were later called by Rutherford. It spontaneously produced heat that could melt its own weight in ice in an hour. It ionized air, it made inert substances fluorescent.

The half-life of radioactive substances was discovered, as it was found that uranium decayed into radium and radium decayed into polonium. Uranium's half-life was calculated to be several billion years, radium's sixteen hundred years, other 'descendants' took only a few hours, minutes, and seconds.

The greatest property of radium, however, was its ability to shrink cancerous tumors. Once this property became known, a 'radium industry' was born. The Curies could have become the richest people on the planet because radium became the most expensive product in the world: it sold for $150,000 a gram. When Pierre asked, hesitantly, if they would want to patent the process, Marie answered "It is impossible. It would be contrary to the scientific spirit".

They gave up their chance at fortune, but the awards and grants started pouring in. After she was awarded her doctorate in 1903, the Royal Institution invited the Curies to lecture on radium. Marie was the first woman ever admitted to that institution. Later that year the Royal society awarded the Curies one of their highest awards: the Davy medal. Pierre, not sure what to do with the heavy gold medal, gave it to Irene to play with this 'big new penny'.

Finally, in December 1903, they received the Nobel Prize for their discovery of radioactivity, to share with Becquerel. With their share of the prize money: 70,000 gold francs, they could finally move out of their leaking shed, and Pierre could stop his arduous teaching jobs and immerse himself in research.

The rest of the Curie saga is well known. Pierre got a professorship in Sorbonne but died a few years later in a traffic accident. Marie, the first woman in France to teach in a French University, was appointed to the same position in Sorbonne that had been held by her husband, and she continued her research till the end of her life. In 1911 she was awarded a second Nobel Prize in chemistry, the only person to that date to win two Nobel Prizes.

She played an important role during World War One, driving an ambulance to the front, with X-Ray equipment in the back, to help surgeons operate on wounded French soldiers by finding shrapnel fragments and bullets under X-Ray examination. She died of leukemia she contracted over the years of exposure to radiation. She was the greatest female physical scientist the world has ever known.

Ernest Rutherford (1871 - 1937)

Reading Rutherford's biography reminds me of Michael Faraday. They both rose from humble backgrounds, had some help and a few lucky breaks in life, and, eventually, accomplished giant steps in understanding the physical universe. They both had powerful personalities and are often referred to as "a force of Nature". It seems that talent, a bit of luck, and a powerful personality are needed to overcome initial handicaps in life.

Rutherford was born in 1871 in New Zealand to a pioneer family with 12 children, a schoolmistress mother, and a wheelwright father. New Zealand at the time was as rural as it gets and the little village where they lived had about 60 scattered houses, a church, a schoolhouse, and a post office.

His early education started in Foxhill school, a few miles from their house in Spring Grove. He was an outstanding student from the start, excelling in every subject he was taught, especially mathematics and science. He never needed to study at home because he instantly absorbed every subject at school. This was very lucky for him because after school he was busy with chores his family needed to be done: chopping wood, felling trees, clearing bush on their land. This early upbringing stayed with him the rest of his life in a somewhat rough and often clumsy demeanor, loud booming voice, and a strong physical presence.

In 1886 he applied for a scholarship to Nelson College and won it with 580 marks out of a max 600 and a perfect 200 score in the arithmetic paper. The scholarship carried free tuition and a 40-pound grant for the residential board. At the age of 16, his father took him over the mountain ranges on horseback. Foxhill was 15 miles distance from Nelson, as the crow flies, but the winding mountain roads made it a much longer journey. His father couldn't support him financially but did everything he could to help his son get an education.

Nelson, New Zealand was a small port town with a small population (even now only 43,000). The college at Rutherford's time had only 80 students, so it wasn't that hard to be an outstanding student and Ernest surely was one. He won prizes in every one of his subjects: English, French, Latin, History and, especially, Mathematics.

He spent three years at Nelson, after which he applied for a scholarship to Canterbury College at Christchurch, 260 miles south of Nelson. He had no problem qualifying for the scholarship and during the next five years (1889-1894) he obtained his BA and M.A. with First Class Honours in Physical Sciences and Mathematics.

During his stay in Christchurch, he fell in love with Mary Georgina Newton, the daughter of his landlady, and they were officially engaged. However, marriage had to wait until Ernest obtained his first professorship. It took him six years and Mary faithfully waited for him in Christchurch while he continued his education at Cambridge University, England.

This is the point at which Rutherford was helped by pure luck, not by his own achievements. Cambridge had just adopted a rule that allowed students from the colonies and Rutherford was the first to qualify. And that's where lady luck came in: Ernest wasn't the first choice for Cambridge for the single position available, but a chemist J. S. MacLaurin. Rutherford resigned himself to get a teaching job in a boy's high school, hoping to apply for the scholarship again as soon as possible. This, however, wasn't necessary, because MacLaurin decided to stay in New Zealand and get married, so the scholarship was transferred to Rutherford in 1895.

As Isaac Asimov reports in the *Biographical Encyclopedia of Science and Technology* "The news reached him, it is said, while he was digging potatoes on his father's farm. He flung down his spade, saying "that's the last potato I'll dig" postponed his own marriage and left for England."

It's not widely known that Rutherford had already started serious and original research in the forefront of science, even before he went to Cambridge: on the topic of generating and detecting radio waves (Hertzian waves as they were called at the time, after the German scientist who first discovered them). He was successful in detecting these waves from distances as much as 60 feet, across a large building. He had published 3 scientific papers on the topic, still in New Zealand, and it was precisely this fact that earned him a scholarship to Cambridge. In this area, he, as a student, was way ahead of Marconi and anyone else experimenting with radio waves at the time.

He arrived in Cambridge in 1895 and he continued this research, as a student of J.J.Thomson, the discoverer of the electron. He increased the accuracy of his detecting method, to be able to detect the radio waves at greater and greater distances, even when whole buildings blocked the line of sight between his transmitter and receiver. As he writes in one of his letters: "I set up the vibrator in the Prof's room at the Cavendish and my detector 100 yards away in Prof Ewing's lab and got quite a large effect through the distance, traversing three thick walls by the way....The next day I tried and got an effect from the lab to Towsend's diggings, a distance over half a mile through solid stone houses all the way."

Even though he got on very well with Thomson, he had to deal with snobbishness from many, as demonstrated by one of his letters, commenting that he had to overcome "ancient prejudices that no good things can come from the colonies". In another letter, he states that "There is one demonstrator on whose chest I should like to dance a Maori war-dance."

A colleague, Sir Henry Dale, admits that "I think that we young cynics might have described his manner as rather hearty and even a little boisterous, but I can imagine us allowing, with pride in our tolerance, that a man who had been reared on a farm somewhere on the outer fringes of the British Empire, might naturally be like that, and might nevertheless have the remarkable ability which rumor was beginning to attribute to him."

Then, when he was way ahead of everyone else in the research of radio waves, his interest changed into a more theoretically exciting area: radioactivity and x-rays. Had he stayed with radio waves, he could have become a very rich man, beating all the others in the field, Marconi in particular. However, he was not commercially motivated, his curiosity trumped every other consideration.

His mentor, J.J.Thomson had been fascinated by the recently discovered X-Rays and asked Rutherford to help him research the effect of these rays on gases. Ernest immediately jumped on the new topic and started to experiment in this direction, abandoning his radio wave research, writing to Mary that "I am a little full up of my old subject."

He spent 3 years in Cambridge as JJ's research student, waiting for an opportunity to get a full-time teaching job. This happened in 1898 when he was offered the Professorship at McGill University in Montreal, Canada. After he took up his position, it took him two more years before he felt secure enough to travel back to New Zealand, marry his fiancé, and travel back with her to Montreal.

Rutherford was a born experimenter. Very much like Faraday and not at all like Maxwell or Einstein who performed only thought experiments. His experiments were ingenious in their directness and simplicity and produced a large number of breakthrough results. Studying the recently discovered radioactivity he quickly identified three types of radiation emitted from the source and he called these alpha, beta, and gamma, for lack of better words. Subjecting each to magnetic fields he discovered that the alpha particles were positively charged and a lot more massive than the beta particles that were negatively charged. This was easy to determine because they were deflected in opposite directions, and at different angles, by the applied magnetic field. The gamma radiation was not affected at all, so they were not charged and, as later determined, they were not particles at all but electromagnetic radiation, similar to x-rays.

His next startling discovery was that alchemists were right after all. Radioactive chemical elements (uranium, thorium, actinium) can undergo spontaneous transmutation from one kind of atom to another. The resulting atoms were not stable but continued to disintegrate into further intermediate states that all had different levels of radioactivity. Thus, he identified a "uranium-series"; a "thorium-series" and an "actinium-series". The rate of the change in each stage was characterized by what he called the "half-life", that is the time it took the radioactivity of the element to fall to half of its original value. They couldn't yet identify the chemical nature of the intermediate stages, it had to wait for further chemical studies to identify each. For example, it was found that uranium has a long chain of transmutations (14 links) via alpha and beta decay that eventually results in the stable element lead (unfortunately, not gold as the alchemists had hoped). One additional hint for what was to come, Rutherford (and his collaborator Soddy) estimated that the energy released from transmutations to a non-radioactive final product was two thousand times the amount produced in even very energetic chemical processes. Their last paper on the subject stated, prophetically, that "the energy latent in the atom …must be enormous compared with that rendered free in an ordinary chemical change."

In Canada, Rutherford felt isolated from the European centers of physics and jumped at the opportunity he was offered in 1907 as Chair of the Physics department at the University of Manchester. He was 36 years old, barely at the start of his career, but his work and publications at McGill firmly established his reputation in scientific circles. His discoveries won him the Nobel prize in chemistry in 1908. That amused him because he considered himself, above everything else, a physicist (a famous quote every biographer must include here: "All science is either physics or stamp collecting") Receiving the Nobel prize before he had achieved his greatest contribution to physics later was quite unusual.

In Manchester Rutherford wasted no time before starting his next research project: to determine the nature and internal structure of the atom. By that time, it was known that the atom had to have an internal structure because that was the only place the electrons, discovered by JJ in 1897, could have come from. What else lurked inside atoms? As he once commented: "I was brought up to look at the atom as a nice hard fellow, red or grey in color, according to taste". For us in the 21st Century, it may seem like a frivolous quest, trying to find out things that we grew up with from kindergarten on, but in Rutherford's time, it was a puzzling mystery: what did the atoms look like close up, under a magnifying glass that had not been invented yet.

Rutherford had one tool to start his investigation with alpha particles (his "strong right arm") emitted from radioactive sources with enormous speed and momentum. He could use them as bullets to bombard atoms of different kinds and see what would happen. However, it took him a major battle with the head of the chemistry department to obtain a minute quantity (500mg) of radium for his own personal use. When he finally received his sample from Vienna, he started working at a frantic pace with his collaborator, Hans Geiger. Their first task was to find out more about these alpha particles. They used two methods of study: the electric method (that later became the famous "Geiger counter") and the "scintillating method" observing the tiny flashes of light produced by single alpha particles hitting a screen coated with zinc sulfide. They had a difficult time filtering out most of the alpha particles, so only a countable small portion would hit the screen. Now that they could count individual particles, they managed to measure the electric charge carried by the alphas and dividing the total charge by the total number, they could calculate the positive charge of an individual alpha particle. The result was exactly twice the already known charge of the electron. This suggested to Rutherford that the alpha particles were Helium atoms stripped of their negative charges.

Proving this beyond any doubt was difficult, but he managed it by the help of an expert glassblower, Baumbach, who produced for him a glass cylinder with walls of less the one-hundredth of a millimeter thick. This sealed cylinder was placed inside another glass cylinder that had taps in its walls, so any gas trapped between the two cylinders could be evacuated. The alpha particles emitted by the radium source placed in the inner cylinder could penetrate the very thin walls, but no other emanation from the radium source could follow it, so the gap between the two cylinders filled up, over time, with pure alpha particles. When enough of these particles accumulated,

Rutherford could detect the spectrum of these particles: the exact spectrum of Helium atoms which was already known.

Having achieved this huge result now he turned his attention to an annoying side effect dogging all his experiments: the slight scattering he observed of a needle-thin alpha particle beam, no matter how carefully he tried to prevent it. As Geiger described the result of his experiments: "the small residuum of gas in the four-meter-long tube…influenced the result. We attributed this to a small scattering of the alpha particles….The most important observation was the appearance of isolated instances of extremely large angles of deflection, which were far outside the normal; variation. At first, we could not understand this at all."

These experiments were carried out in 1909 by a young undergraduate student, Ernest Marsden who, according to Rutherford's later account: "Why not let him see if any alpha-particles can be scattered through a large angle?" The result was unbelievable: a countable number of alpha particles bounced back from a thin sheet of gold foil. The majority of the alphas went straight through the gold foil and were only very slightly deflected. Rutherford described his astonishment (often quoted by biographers) "It was as though you had fired a fifteen-inch shell at a piece of tissue paper, and it bounced back and hit you."

Marsden managed to show that one alpha particle out of 8000 was deflected by more than an angle of 90 degrees. Further experiments and calculations followed at a feverish pace. Actually, Rutherford went back to school: he enrolled as a student to attend the elementary lectures on probability to learn the mathematical tools he needed to calculate the probability of this extraordinary result happening. Further experiments using eight different common metals resulted in deflection increasing with the atomic weight of the metals.

It took Rutherford eighteen months of further experiments and calculations to come up with a working hypothesis: the atom must contain a massive charged center of very minute dimensions, containing most of the mass of the atom, surrounded at great distances by a distribution of electrons. The concept of the nucleus atom was born. This concept was radically different from JJ

Thomson's "plum-pudding" model that envisaged the atom as a solid sphere with the electrons embedded inside it.

Thus the results were published by Geiger and Marsden in 1913 as " an underlying assumption that an atom containing a strong charge at the center, of dimensions, small compared to the diameter of the atom". Rutherford's calculations showed that the diameter of the massive, charged nucleus was 10^{-18} centimeters. Compared to the atom's total diameter of 10^{-13} centimeters – the atom is mostly empty space, no wonder that most of the alpha particles just sailed right through.

Acceptance of this hypothesis was slow in coming until the physics community realized its extraordinarily successful explanatory power when applied to mysterious phenomena such as anomalies in Mendeleev's table and spectra.

The first hint about more mysteries was the discovery by Soddy that in the radioactive chains one could find elements with identical chemical properties, yet with different atomic weights. Soddy called such pairs "isotopes" – occupying the same position in Mendeleev's Table. This was difficult to explain at the time.

Now that the atom was proven to have an internal structure, further research discovered that the nucleus itself also did have an internal structure. Alice's rabbit hole was leading the unsuspecting physicists deeper and deeper.

The next step in discovering the nuclei's internal structure came from Rutherford's experiment of bombarding hydrogen gas with alpha particles. He discovered the hydrogen nucleus and started calling it "proton" – and this proton was more energetic than the alphas. Using nitrogen gas instead of hydrogen, again he discovered that hydrogen atoms were liberated as a result of a collision with alphas. This proved to him that the nitrogen atom *disintegrates* due to the bombardment with the high-energy alphas. Atom smashing had started and is still going on strongly today.

In 1919 Rutherford left Manchester to take up the position of Chair of the Cavendish Laboratory in Cambridge. There he started his next intensive research into the internal structure of atomic nuclei. He worked very closely with James Chadwick and Charles Wilson, the inventor of the "cloud chamber" that enabled physicists to visually follow the path of individual charged particles (alpha, beta, proton, and electron) as they ionized the water vapor molecules in their path. Observing the paths, they discovered some 'forked' ones proving, visually, that atoms disintegrated as a result of the collision. This was the beginning of cyclotron accelerator experiments by Ernest Lawrence in California.

Now Rutherford had all the tools he needed to start his last momentous discovery of what he called 'neutrons' (experimentally proven by Chadwick in 1931) by suggesting the existence of an electrically neutral particle that could be obtained by combining a proton with an electron. This neutron would explain the atomic weight of the heavier elements. He summarized his theory by suggesting that the number of protons in a nucleus determines its positive charge and the number of protons and neutrons its atomic weight.

Rutherford had gone a long way from the "the atom as a nice hard fellow, red or grey in color, according to taste."

James Chadwick (1891-1974)

The most prophetic thinkers of any age are usually science fiction writers. Among these visionaries, H.G. Wells often proved to be eerily accurate in his predictions. In his novel "The World Set Free", published in 1914, he predicted atomic bombs, nuclear war, and world government. We are still waiting for the last part, but the rest is now in our history. The novel starts with a penetrating insight: "The history of mankind is the history of the attainment of external power. Man is the tool-using, fire-making animal. . . . Always down a lengthening record, save for a set-back ever and again, he is doing more."

The saga of the road to this destructive power started, as often is the case, by physicists. Once Rutherford experimentally proved the existence of atomic nucleus, nuclear physics took off in a big way. Lise Meitner in Germany, the Joliot-Curies (Marie's daughter and husband) in France, Enrico Fermi in Italy, and, most of all, James Chadwick in England contributed heavily to this new science that resulted in peaceful (although dangerous) nuclear reactors and deadly nuclear weapons. We are still saddled with the consequences of this major advancement in the understanding of the physical world.

The ball started rolling by Rutherford's 'Bakerian Lecture' on June 3 1920 before the Royal Society of London. He speculated about a possible nuclear constituent with a mass of one and zero charge – a combination of a proton and an electron very closely attached to each other. If such a 'doublet' did exist, "it would have very novel properties" as Rutherford explained. He suggested that it would be very difficult to detect because of its electric neutrality and might be very difficult to contain. "on the other hand, it should enter readily the structure of atoms, and may either unite with the nucleus or be disintegrated by its intense field". He assumed that if such a particle did exist (he called it a *neutron*) it might be the most effective of all tools to probe the atomic nucleus. Prophetic words, indeed.

Rutheford's young assistance, James Chadwick, who was present during this lecture, was fascinated by this possibility and spent the next 12 years of his career finding this elusive particle.

Chadwick was a shy and timid boy at 16 when he won two scholarships to the University of Manchester. He intended to be a mathematician, but he joined the wrong line of students lined up for the test and, during his examination, he realized that he was tested for the Physics course. He was too embarrassed to admit his mistake, so he went through with the exam and got accepted to start his education in physics. It was during his second year when he heard a Rutherford lecture that he converted to his unintentional choice for his future profession. He graduated in 1911 with first-class honors. In 1913 he won a research scholarship that required him to move to Berlin. Bad choice because WWI started soon after he got there, and he found himself in an internment camp with other British nationals. After the war he was back in Marchester, working in Rutherford's laboratory, looking for the elusive neutron. He never suspected that it would take him 12 years to find the damn thing.

Following Rutherford's guess of the neutron being a close marriage between a proton and an electron, he started by subjecting the hydrogen atom to electric discharges, hoping that somehow they could unite the hydrogen nucleus (a proton) with an electron. No such luck, the hydrogen nucleus remained stubbornly stable. His experiments required more precise methods of detecting radiation than watching and counting for scintillation on the detecting screen, so his collaborator Hans Geiger invented his electrical counter, now known as the Geiger counter, that could be used more precisely and easily than the scintillation method.

Finally, in 1928, following a German physicist's, Walther Bothe's, lead they started bombarding light elements, such as boron, magnesium, and aluminum, with alpha particles, to see what would happen to their nuclei. They were interested in the gamma radiation that accompanied the excitation by the alphas but found, to their surprise, that with beryllium the emitted radiation was almost ten times higher than with any of the other light elements. This was strange and needed further study. That was even stranger, the radiation excited from the beryllium target had more energy than the total energy of the bombarding alphas, even though no protons were emitted together with the gamma radiation. Where did the extra energy come from?

Chadwick's research student, an Australian by the name of H.C.Webster, who studied this strange phenomenon further, discovered another strange thing: "the radiation from beryllium which was emitted in the same direction as the … alpha particles were more penetrating than the radiation emitted in the backward direction." This didn't make any sense because the gamma radiation should have been spherically symmetrical. Nothing could explain the energy difference …unless…a new and *neutral* particle was also emitted from the beryllium nucleus, dislodged by elastic collision from the incoming alphas, bumping it forward.

Chadwick had no doubt: "and that, of course,… was a point which excited me very much indeed because I thought, 'Here is the neutron'." In the meantime, in Paris, the Joliot-Curie pair studying the same phenomenon was still convinced that the radiation with the extra energy emitted from the beryllium atom was gamma rays. The further experiment proved that this highly energetic radiation from the bombarded beryllium ejected high-velocity protons from hydrogenous materials like paraffin wax. The Curies still tried to explain this by assuming the well-known Compton Effect that could deflect electrons, but protons were 1836 times heavier than electrons. Chadwick didn't believe this for a second. He was sure: this was the neutron he had been searching for, for 12 years.

At this point he took over from Webster, fully determined to personally conduct the last phase of his project. The first result he obtained was that the beryllium radiation could pass through a two-centimeter-thick block of lead. Testing the power of the radiation with different substances he found that it could knock protons out of every material he tried. Not only that, but the energies of the knocked-out protons were far greater than what could be explained by the gamma-ray photons the Curies assumed.

In Chadwick's words: "" If we suppose that the radiation is not a [gamma] radiation, but consists of particles of mass very nearly equal to the mass of the proton, all the difficulties connected with the collision disappear, both with regard to their frequency and to energy transfer to different masses. In order to explain the great penetrating power of the radiation, we must further assume that the particle has no net charge…We may suppose it [to be] the 'neutron' discussed by Rutherford in his Bakerian lecture of 1920."

In his announcement to the members of a distinguished club, the "Kapitza Club" Chadwick gave full credit to Bothe, Becker, Webster, and the Joliot Curies. He finished his brief announcement by saying: "Now I want to be chloroformed and put to bed." He has spent the previous three weeks working with no more than 3 hours of sleep a night.

The discovery of the neutron explained the weight of the elements without adding an extra charge. It also explained why it was nearly impossible to separate isotopes that had the same number of electrons (responsible for their physical and chemical properties) but different atomic weights – they differed only in the number of neutrons inside their nucleus.

Also, because the neutron had no electric charge and could not be deflected by the electron shell of atoms it was aimed at, it had great penetrating power that could be used to further study the structure of the atoms. Pandora's box was open, and it did not take long for other physicists to peer inside it.

Chadwick received the Nobel Prize in 1935 and he was knighted in Britain in 1945.

<u>Lise Meitner (1878 – 1968)</u>

Lise Meitner's story is a demonstration of how talent, brilliance, and endurance can overcome the most blatant form of discrimination. She wasn't the only woman affected by misogyny: science history is full of it.

Henrietta Swan Leavitt, Edwin Hubble's assistant, began working in astronomy as an *unpaid* volunteer, forbidden to touch the telescopes. Her brilliant insight led to our modern understanding of the true size of the universe. Hubble used Leavitt's period-luminosity relation, in order to establish that the universe is expanding - his most famous contribution to science. The Swedish Academy of Sciences tried to nominate her for the Nobel prize in 1925, only to learn that she had died of cancer three years earlier. The irony! Hubble's name is known all around the world, Leavitt died in obscurity.

At the end of the 19th century, public education in Austria for girls ended at age 14 and, unless your family could afford to pay for private tutors, women couldn't even dream of university education. Lise Meitner, like Marie Curie, mostly educated herself and successfully passed the entrance examination to the University of Vienna (she wasn't outright forbidden to enter like Maria Sklodowska was in Poland) and became an enthusiastic student of Ludwig Boltzmann. She got her doctorate in 1905 with summa cum laude – the second woman to earn a doctorate at that university. In 1907 she decided to go to Germany to work with Max Planck.

He was kind but did not have a very high opinion of female students. As she recalled later, Planck asked her about her plans in Physics: "But you are a Doctor already! What more do you want?"

(This attitude is universal as far as I can tell – at my own university in Hungary I had a professor who was famous for telling students, in front of a class of 150 mixed-sex audience, that "during exams, I ask this idiotic question only from girly students")

At Planck's house parties where Meitner was often invited, she met Otto Hahn, a radiochemist who became her lifelong friend and collaborator. Hahn on the chemistry side, Meitner on the physics aspects of the research they did together. They were of the same age, but Hahn was already an assistant in the Berlin Chemistry Institute, while Meitner wasn't even allowed to set foot inside and had to work in the basement in a carpenter's shop with an outside entrance. She had to walk to a corner restaurant to use the toilet. Misogyny on steroids indeed. Her status changed somewhat in 1909 when she was finally allowed in the laboratories as an unpaid assistant. During this time Meitner and Hahn published three major articles together.

In 1912 the Meitner-Hahn duo moved up another notch in their standing. At least Hahn did. He was appointed as "scientific associate" of the Kaiser Wilhelm Institute of Physical Chemistry. Meitner was *allowed* to work with him as an unpaid 'guest physicist". Her first paid academic position was a part-time job as Planck's assistant – doing menial work grading student papers that Planck was too busy to bother with. Finally, a few years later Meitner got her first academic title as "scientific associate" at a substantially lower salary than what Hahn received.

Meitner's main interest in Physics was radioactivity and, within that topic, beta radiation in particular. Ever since she had read about Marie Curie's discovery of the phenomenon she was fascinated by the mystery of where the radiation came from. As Rutherford had pointed out, two types of particles were emitted during radioactive decay: alpha and beta particles. The alphas were identified as positively charged helium nuclei, while beta particles turned out to be high-energy electrons. However, alpha particles always appeared with roughly the same energy, while beta particles can have a wide range of energies from almost zero to a certain maximum value. Why the difference? Where do these electrons come from? By that time the model of the Bohr atom was widely known, according to which the nucleus was the positively charged center, while the electrons were distributed on the outer shell of the atom.

Meitner believed that the electrons were emitted from inside the nucleus where they did not exist prior to the emission, but were 'created' by the disintegration process that turned the atom's nucleus into the nucleus of another element, as it was discovered by Frederick Soddy, Kasimir Fajans and A.S.Russell in 1912. They had found that when an element emits a beta particle, which is a high-energy electron, the element is transformed into the element with the chemical properties of the next element above it in the periodic table. The mystery was in the energies: The energy equation was not balanced for the process. The total energy of the released beta particles and the newly created nucleus did not add up to what it should have. So where did the missing energy disappear to? Bohr, the epicyclist, was even tempted to suggest abandoning the conservation of energy principle for atomic systems. Wolfgang Pauli saved physics from that ignoble fate: he suggested that a new particle he called 'neutrino' was created in the process and it carried away the extra energy. A complete explanation had to wait for Chadwick's discovery of the neutron after which Enrico Fermi gave a full description of the beta decay process.

In the meantime, Meitner and Hahn continued their collaboration until 1938 when Hitler's anti-Semitic laws made it impossible for Meitner to remain in Germany. For a while, she could hide behind her Austrian citizenship but after the Anschluss, she had to flee.

However, she could not travel under Nazi rules and she did not even have a valid passport anymore. One of her many scientist friends, Dick Coster, traveled with her to the Dutch border and bribed some border guards to let her slip over to Holland. She was out of Germany, destitute because all her assets were frozen in Germany, but not quite safe yet. She saw the writing on the wall about the future Nazi invasion of Holland, so she accepted an offer from Sweden and arrived in Stockholm on Aug 1, 1938. Her exile from Germany has begun.

The breakthrough in her research came before Christmas, 1938 when she went skiing with her physicist nephew, Otto Frisch, in a Swedish forest. She was reading a letter she received from Hahn, in which he reported a mysterious result of a chemical experiment: after bombarding uranium atoms with neutrons. He found traces of barium, a much lighter element, with the atomic weight of about half of uranium. He couldn't trust his own experiment because, as he wrote "We know ourselves that [uranium] can't actually burst apart into [barium]…If there is anything you can propose that you could publish…you will do a good thing if you can find a way out of this."

When Hahn published his own work, he was cautious interpreting his own results. "…as nuclear chemists close to physics, we cannot yet take this leap which is contrary to all experience of nuclear physics." Meitner, however, was ready to take this leap. In her nephew's words:

"…we both sat down on a tree trunk and started to calculate on a scrap of paper. The charge of the uranium nucleus, we found, was indeed large enough to overcome the effect of the surface tension almost completely; so the uranium nucleus might indeed resemble a very wobbly, unstable drop, ready to divide itself at the slightest provocation, such as the impact of a single neutron."

They realized that the two newly created positively charged nuclei would repel each other with tremendous energy. Where was that energy to come from? Then she could see it. The mass difference between the uranium nucleus, and the combined masses of the fragmentary nuclei, converted by Einstein's famous equivalence formula ($E=mc^2$), would be exactly the explosive energy of the resultant light weight nuclei.

The destruction of Hiroshima and Nagasaki was foreshadowed on that fateful day in a snowy Swedish forest.

In 1944 Otto Hahn received the Nobel prize in chemistry, while Lise Meitner was ignored by the Swedish Nobel committee. So what? She was *only* a woman.

Qantum Physics - "Shut up and calculate!"

Everybody is familiar with the Indian parable of the three blind men trying to describe an elephant. The first blind man touches the elephant's trunk and says, "An elephant is like a snake." The second blind man, while touching the elephant's leg, says, " An elephant is like a great trunk of a tree." The third blind man, touching the side of the massive animal, says, "An elephant is like a brick wall." Each man thinks he is right, and the others are wrong. That's what Quantum Physics has become during the last 100 years.

The most incisive physicist of our time, Richard Feynman, said it eloquently in his *Lecture on Physics* textbook: "We choose to examine a phenomenon which is impossible, *absolutely* impossible, to explain in any classical way, and which has in it the heart of quantum mechanics. In reality, it contains the *only* mystery. We cannot make the mystery go away by "explaining" how it works. We will just *tell* you how it works. In telling you how it works we will have told you about the basic peculiarities of all quantum mechanics."

What Feynman meant by "classical way" is the only way we have experienced the world up to this point, the only way we can understand cause-and-effect relationships.

At this point, Physics entered a crisis period that it still has not recovered from. A large number of books have been written about the crisis in Physics. Just to mention a few:

Nick Herbert: "Quantum Reality" (1985)
David Lindley: The End of Physics (1993)
John Horgan: "The End of Science" (1997)
Helge Kragh: "The Quantum Generations". (1999)
Lee Smolin: "Trouble with Physics" (2006)
Sheilla Jones: "The Quantum Ten" (2008)
Alexander Unzicker and Sheilla Jones: " Bankrupting Physics" (2013)

Quantum mysteries

Let's start at the beginning, with simple experiments strongly grounded in rock-solid reality.

<u>Mystery #1: The Airy Pattern</u>

When we aim an electron gun at a surface with one small hole in the middle, we see a spot of light appear on a phosphor-screen placed behind the hole. If we shrink the size of the hole gradually to smaller and smaller, at some point the spot of light on our screen is replaced by concentric circles of bright and dark rings, like in an archery target. If we keep on shrinking the size of the hole, the image on the screen gets larger and larger until it fills the entire screen. It was George Biddell Airy in 1835 who first demonstrated this phenomenon. As he explained, these rings appear any time you aim any kind of wave (sound, water, light) at a small hole. He didn't yet know about electrons and electron guns, so he used an ordinary light source).

Next, we reduce the intensity of the electron beam to the point where only one electron passes through the hole each minute (on average). On the screen, we see random flashes in no recognizable pattern. G.I. Taylor was curious to see what would happen in the long run. He replaced the phosphor screen with a photographic plate and went away on a sailing trip. When he returned a few weeks later and developed the film, he saw the same pattern of concentric rings appear.

The thousands of dots, created by single electrons, going through the hole one a minute, weren't randomly distributed but reproduced the same Airy pattern. How does a single electron, on its own (without interacting with anything else), 'know' how to find its place in the Airy ring?

<u>Mystery #2: The double-slit experiment.</u>

The experimental setup is almost the same as with Mystery #1, except we have two small slits on the barrier between the electron gun and the phosphor screen. If we close one of the slits and turn on the electron gun, we see a clear glow on the screen as the electrons going through the slit hit the screen. However, if we leave both slits open, instead of the two spots on the screen that we had expected, we see lines of alternating black and white bars on the screen. Clearly, an interference pattern. The assumption: the two-electron beams going through the two holes are interfering with each other like water waves would when going through a barrier with two holes.

Assumption: any one electron goes through either slit #1 or slit #2. Can we find out which hole a single electron goes through? If we reduce the voltage in the electron gun low enough, eventually only one electron at a time goes through the slit. Now we can place a detector (that would produce a flash if an electron goes near it) on the electron gun's side of the barrier, next to each slit. Then, watching for the telltale sign of flash, we will know which slit the electron went through. We will see the flash all right at one of the slits, but a strange thing happens if we let the process go long enough – we no longer have an interference pattern from the accumulated dots on the photographic plate. However, if we turn off the lights and, thus, stop watching and finding out which slit the electrons went through, then the interference pattern returns. Countless experiments with trying to watch the slits ended the same way. If we watch which hole the electron went through, then the interference pattern disappears.

What if the light from our detector disturbs the electrons? We know from the photoelectric effect (see later) that light can knock electrons out of a metal surface. Maybe that's what is happening here to change the pattern? If we turn the light down, dimmer and dimmer, will it make a difference? No such luck. But, wait a minute! We also know that the intensity of the light doesn't affect one electron's energy, only the number of electrons knocked out of the metal But, the wavelength of the light does. So, if we make the light redder and redder (longer and longer wavelength) then we will see a difference. Nothing happens for a while, but, at the point when the wavelength is longer than the distance between the slits, enough to limit the optical resolution to the point where we can't tell the two slits apart, then all we will see is a big fuzzy flash, so we can't tell which slit the electron went through. Then the interference pattern reappears. However, at this point, we can't tell which hole the electron went through. In Richard Feynman's words, "a terrible thing happened". When we can't watch the electrons, they change their behavior. Mystery #2: How does an electron 'know' that it is being watched? How can it 'know' that we can't tell anymore? And why does it 'care'?

Mystery #3: The delayed choice experiment

Originally a thought experiment proposed by Archibald Wheeler (based on the phenomenon of galactic lensing). What if we modify the arrangement in Mystery #2 and place the detectors on the other side of the barrier so that by the time the detector records the electron, it has already passed through the hole. This has to be done with extreme precision (controlled by a computer), based on the speed of the electron and the distance of the detector from the barrier. A source sends electrons one by one. A removable screen is placed between the barrier and the two detectors (each one pointing precisely at one slit). If the screen is present, then after a while, an interference pattern is seen. If the screen is removed, for each electron only one detector clicks allowing us to know which slit the electron went through. However, at this point, the interference pattern disappears from the photographic plate behind the detectors. Somehow the electron 'knows' even after it has gone through one or both slits and has no way to interfere even with itself while passing through.

A possible interpretation would be that the electron is "informed" of the presence or of the absence of the screen (no matter how far away it is) when it passes the slits. A delayed choice situation corresponds then to the case where choosing to put or remove the screen is done after the electron went through the slits. In this case, the conclusion seems to be that putting or removing the screen has an influence in the past on the 'decision' of the electron to produce or not an interference pattern.

Mystery #4: Quantum Entanglement:

The most mysterious behavior of electrons (and other quantum entities like protons, neutrons, photons, etc.) is what is called entanglement. "Quantum entanglement means that you can't describe your joint quantum system in terms of just local descriptions, one for each system," says Michael Hall, a theoretical physicist at the Australian National University.

Entanglement first arose in a thought experiment worked out by none other than Albert Einstein. In a 1935 paper, Einstein and two colleagues showed that if quantum mechanics fully described reality, then conducting a measurement on one part of an entangled system would instantaneously affect our knowledge about future measurements on the other part, appearing to send information faster than the speed of light, which is impossible according to all known physics. Einstein called this effect "spooky action at a distance," implying something fundamentally wrong with the budding science of quantum mechanics.

Decades later, quantum entanglement has been experimentally confirmed time and again. Actually, current designs of quantum computers use the phenomenon of entanglement. While physicists have learned to control and study quantum entanglement, they've yet to find a mechanism to explain it or to reach a consensus on what it means about the nature of reality. "Entanglement itself has been verified over many, many decades," says Andrew Friedman, an astrophysicist at the University of California, San Diego, who worked on the quasar experiment, also known as a "cosmic Bell test." "The real challenge is that even though we know it's an experimental reality, we don't have a compelling story of how it actually works."

Quantum Realities

Before you rejoice at the prospect of a breakthrough in modern physics: an answer to the mysteries I told you before, I'll have to disappoint you: there isn't any. More precisely: there are two lines of attack physicists attempted over the last century: one spectacularly successful and the other confusing at best, pure speculation at worst.

The first line of attack was what the title of this book is trying to describe: epicycle physics. This title refers to the highly successful mathematical theory of Ptolemy, which predicted the position of the planets with remarkable accuracy. The underlying theory he based his calculations on was the assumption that the planets move around a point on a big circle around the Earth, while also moving around a smaller circle (the epicycle) around that point on the big circle. This complicated motion was necessary to explain the planets' irregular, retrograde movement that they had been observing for centuries. After tweaking his assumptions about these hypothetical epicycles (the number of them, their radiuses, etc.) he managed to fit his theory to the observed facts. In other words: he cheated. That was not unusual in science then or even now, nor is it particularly unethical: scientists always make assumptions and then test the predictions that follow these assumptions. However, usually, they have some reason why they make these assumptions, other than to come up with the desirable result.

Sometimes, even world-class scientists like James Klerk Maxwell indulged in dreaming up clever contraptions and rules that these contraptions have to follow to simulate the behavior of real physical systems. Without a shred of experimental data, Maxwell imagined an elaborate mechanical clockwork model with invisible spinning cells and rotating, moving wheels permeating all space. Yet, when he developed his mathematical equations, they described the magnetic Coulomb Law, the magnetic induction effect of an electric current, and the induction of an electric current by a changing magnetic flux. He fully acknowledged that his contraption was a pure thought experiment, nothing to do with reality. Still, it was Epicycle Physics.

Jumping forward from Maxwell to the twentieth Century, that's what physicists attempted to do: Make some (unwarranted) assumptions and develop mathematical theories that could predict the likely results of future experiments. These mathematical theories, which are collectively called "Quantum Mechanics" worked spectacularly well and made predictions that turned out unerringly accurate. However, the assumptions were totally arbitrary and unintuitive – no human mind managed to justify them in any comprehensible way during the last and even in the present century. What was even worse, some of the assumptions contradict laws established in classical physics.

In the second line of attack so far at least eight major philosophical interpretations were proposed by world-class physicists and scientists. Nick Herbert summarized these interpretations in his excellent book: *Quantum Reality*. Before I list them here, I have to state my own personal opinion: these 'interpretations' are pure philosophical speculations and, as such, they are not real physics. They all ponder the nature of 'reality' because, I presume, quantum facts seemed so unreal to them that they were forced to examine this nagging question: what is, and what isn't, real? What do we mean by real? The first of the eight was proposed by Niels Bohr, that's why it is called "The Copenhagen Interpretation" (Niels was Danish and had his institute in Copenhagen)

#1 There is no deep reality.

This outrageous interpretation was suggested by Niels Bohr and Werner Heisenberg and it is now the mainstream interpretation by the physics community. What did they mean by reality? Did they deny the evidence of their senses? No, they admitted it seems real enough but it 'floats' on a world that is not as real. This world is not real as it exists objectively because "there is no quantum world only quantum description". Heisenberg wrote: "The hope that new experiments will lead us back to objective events in time and space is as well-founded as the hope of discovering the end of the world in the unexplored regions of the Antarctic. Another physicist summed up Bohr's position by saying "we now know that the moon is demonstrably not there when nobody looks".

#2 Reality is created by observation

Another variation of interpretation #1: There is no 'deep reality', only "phenomenal reality" – what we see is undoubtedly real, but it doesn't exist in the absence of observation: observation creates reality. Proponents of #2 can't agree among themselves about crucial questions like "which of the world's myriad processes qualify as observations?"

#3 Reality is an undivided wholeness

Everything is connected below the surface to a seamless whole. There is no separation between an objective outside reality and the conscious observer, they are one and the same. David Bohm wrote: "One is led to the notion of unbroken wholeness…the inseparable quantum interconnectedness of the world is the fundamental reality".

#4 Many world interpretation

"Reality consists of a steadily increasing number of parallel universes". On each measurement, however made, a new parallel world is created. For each observation for which different results are possible, a new universe is created where the measurement result is different accordingly. Science fiction writers have been using this notion to come up with plots requiring parallel universes.

#5 The world obeys non-human reasoning

Just as Einstein abandoned Euclidean geometry in favor of a crazy, non-intuitive one (Riemann's) in order to develop his general theory of Relativity, proponents of this interpretation say we need to do the same with logic: abandon the Boolean logic we have been using for two thousand years and develop a new "quantum logic" to fit the observed facts of quantum experiments. I am not sure how this could be accomplished because, in my mind, logic mirrors the way our brains work.

#6 Neorealism (The world is made of ordinary objects)

Ordinary objects are real whether anyone observes them or not. As Einstein wrote: "The Heisenberg-Bohr tranquilizing philosophy – or religion? – is so delicately contrived that for the time being, it provides a gentle pillow for the true believer from which he cannot very easily be aroused". Proponents of the neorealist interpretation include Erwin Schrodinger, Max Planck, and Louis de Broglie.

#7 Consciousness creates reality

This is the silliest idea so far because it begs the question of what creates consciousness. Chicken and egg, as usual. Still, world-class mathematician John von Neumann and Nobel laureate Eugene Wigner (both Hungarians) signed up to this self-centered idea, reminiscent of the Earth-centered philosophy of the middle ages. We are special, it's all about us, the world revolves (or is created) around our human consciousness.

#8 The world consists of potentials and actualities

This interpretation posits that phenomena aren't real, only 'potentialities' are. Have we not heard this before? One can't help thinking about Plato and his 'world of ideals" that are the only real things and we only see their reflections on the cave wall. Nothing is new under the sun.

My own, personal take on these lofty philosophical interpretations is that, in order to live our lives, we need to (explicitly or implicitly) assume the existence of objective reality. As far as the conundrums and illogical, contradictory, and plain impossible behavior of quantum particles are concerned, I humbly accept the possibility that the human brain (mine included) is not equipped to understand the phenomena. One hundred and twenty years of failed attempts by thousands of very bright individuals may be a strong hint that this is the case. We have evolved as survival machines in the macro world and our brains developed the necessary circuitry to deal with this world. Maybe we just don't have what it takes to go beyond that. This does not mean, obviously, that we should stop trying, but we always need to know what it is that we know, how do we know it and what are the limits of our understanding. I find this possibility strangely comforting because I find none of the eight interpretations even remotely satisfactory. That may be because I am not a genius myself, but I am a reasonably intelligent person with years of training in physics. I always insist on the precise definition of words and concepts and I have not found them in the philosophical literature.

For example, what does the word 'objective' mean? What can it mean? As long as I am alive, looking at the world from inside my own mind, everything is subjective, by definition. So how can I talk about objective anything when everything I have experienced and ever will experience is necessarily subjective? Do I mean that the moon is still there even if no one is looking (directly or indirectly)? How can I test that theory? In one of my favorite movies, a philosopher is trying to impress a pretty young waitress, telling her that she can't prove that she exists. Her answer: "Why don't you assume that you don't exist and see where that assumption takes you." I am deeply suspicious of philosophical arguments and this is one thing I have in common with Richard Feynman.

If you feel that these interpretations (except for mine, of course) represent deeply disturbed minds, remember that they were proposed by genius caliber scientists. Also, remember that they were prompted by the incomprehensible experimental facts that started quantum theory in the first place. The human mind can't handle the logical contradiction apparent in the behavior of the very small. Our brains are not wired this way, our logic cannot tolerate contemplating the logically impossible (hence the quantum logic interpretation).

Earlier, I stated that "Contradiction in science is like poison in the human body – it has to be expelled for the body to survive. "This is not true in Epicycle Physics anymore, as you'll soon see. So where did it all get started? Who was the first Quantum Epicyclist who made the first step? It was the most unlikely physicist, Max Planck, a "reluctant revolutionary" as he was often referred to. He was the first who, out of pure desperation, invented the word and concept of 'quantum'.

Max Planck (1858 - 1947)

In the words of his biographer, J.L. Heilbron, "Planck was not a genius. His teachers at the Maximilians-Gymnasium in Munich ranked him near, but never at, the top of his class". What he had was character and a stubborn belief in the world being ordered, absolute, and independent of human perception. He also believed that Physics was the discipline to understand this implicit order and he was fully determined to contribute to this understanding to the best of his ability. Even though at 17 he was advised not to choose Physics as his future profession, because all the significant laws – he was told - have been already discovered, he ignored the spectacularly naïve advice and immersed himself in his chosen subject, looking for that 'absolute truth' he fervently believed in.

He was convinced that the laws of classical physics and, within that, Thermodynamics was the key to the absolute principles governing the universe since Thermodynamics deals with the most fundamental concepts of 'energy' and the recently discovered 'entropy'. All the processes observed in nature, both inanimate and biological have to do with some form of energy transfer and, without exaggeration, we can consider it a 'life and death' question. If and when energy stops flowing, the universe as we know it is dead. (I can sympathize with this fascination because I chose the same topic for my graduate degree thesis in theoretical physics for similar reasons.)

Planck's main project in pursuit of discovering the fundamental laws of Thermodynamics was studying what was known as the "Black body radiation". We know from electrodynamics that both the color (light) and heat of objects represent electrodynamic radiation of different frequencies. Blue color represents higher, red color represents lower frequencies. Frequencies of EM radiation below the human eye's ability to detect are called infrared radiation (below red). On the other end of the spectrum, EM radiation with frequencies too high for the human eye is called ultraviolet radiation. All of these EM radiations can occur at a different intensity, therefore emitting different levels of energy.

It is important to realize that heat and light are closely related. Heated objects, like a stove or a desk lamp's filament, emit electrodynamic radiation, part of a continuous electromagnetic spectrum. Visible light is only a segment of the spectrum that the human eye can detect. Some animals and insects can 'see' in a different range of the EM spectrum than humans. The lower frequencies (infrared) are invisible for us but we register it as heat.

Objects not only emit EM radiation; they absorb it too. That's why we get warm and then hot, sitting in the sun. Our body absorbs the energy of the EM radiation of the sun and heats up. So, any object constantly and simultaneously both absorbs and emits EM radiation. If it is in a state of equilibrium, it emits exactly the same amount of energy as it absorbs. If not, then it keeps cooling down or heating up according to the difference between what it absorbs and emits, until it reaches a state of equilibrium.

A black body is black because it absorbs all radiation falling on it, without reflecting any of it back. Remembering the "emission spectrum" and the "absorption spectrum" described earlier, we know that hot hydrogen gas radiated the same frequencies as the cold hydrogen gas absorbed. It was natural to assume that a black body that absorbed all frequencies when cold would radiate all frequencies when hot. The hotter the object the more energy it radiates. The question arose: how is the energy distributed on the different frequencies emitted? Experiments showed that the energy varied with frequency in a rising curve that reached a maximum at a certain frequency and then steadily decreased again.

The shape of this curve depended to a large extent on the material and the surface of the body, so physicists created a standard test object that could be used for exact measurements. The test object chosen was the inside of a closed oven with a tiny hole drilled in its side. The assumption was that it would approximate a perfect black body that absorbed all frequencies without reflecting any of it back. As they started to heat the oven, energy radiated out through the tiny hole and then they could measure the energy distribution of the radiation of different frequencies coming out of the hole. Since the test object absorbed all frequencies, it would emit all frequencies again, so they could get a continuous curve: an energy level for every frequency on the whole EM spectrum.

Wilhelm Wien (1864 – 1928) German physicist measured this distribution at different temperatures of the oven and found that the peak of the curve shifted to the right (higher frequencies, bluer color) with the rising temperature. That is why we see the color change of heated objects from dull red to bright red to yellow-white to blue-white. This shift of the peak wavelength is now called Wien's displacement law. Wien also produced a mathematical formula that described this displacement of the peak energy depending on the temperature and it worked reasonably well for high-frequency radiation.

$$\lambda_{peak} = b/T$$

where λ_{peak} is the wavelength at which peak energy is radiated, T is the absolute temperature, b is a constant with the value of $b \approx$ 2898 μm·K. Planck decided to derive a formula that would work for all frequencies, better than Wien's. The formula that he came up with worked extremely well, corresponding with all the available measurements, for all frequencies. However, this was a semi-empirical formula and Planck wanted to provide a theoretical framework that could deduce this formula from basic principles.

Since it was known that the energy distribution (over different wavelengths) did not depend on the material of the walls inside the oven, Planck represented these walls as 'resonators', each of which was a massless spring with an electric charge at the end. As the oven was heated, the springs started 'resonating' with different frequencies, depending on the temperature. Now Planck could use the laws of thermodynamics and electrodynamics to deduce his law.

Planck wanted to use the concept of entropy, his favorite subject, to build his theory, in particular using the entropy-energy relationship in thermodynamic systems. For this task he had to rely on Ludwig Boltzman's interpretation of the second law of thermodynamics, according to which the entropy of a system was proportional to the logarithm of the probability of disorder for the state of the system:

In Planck's formulation:

$$S = k \ln W$$

Planck was not very happy with Boltzman's statistical law: he deeply distrusted imprecse, and indeterminate laws that required calculations in probability, which felt like gambling to Planck. However, as Boltzmann's law was his only possible starting point, he reluctantly accepted it.

As he later wrote to a friend: "By nature, I am peacefully inclined and reject all doubtful adventures, but by then I had been wrestling unsuccessfully for six years with this problem of equilibrium between radiation and matter...a theoretical interpretation had to be found at any cost, no matter how high."

The cost turned out to be very high indeed: the only way Planck could make his equations produce results that agreed with measurements was by making a totally arbitrary assumption, a mathematical trick, that required that the total energy of the resonators could increase only in discrete lumps, proportional to the frequency of the resonators.

$$\varepsilon = h\nu$$

With 'h' as the proportionality constant (now called Planck's constant) and 'v' is the frequency of the resonators. With this assumption, he could derive his empirical formula from basic principles. He hated it, called it a 'capitulation', and was convinced that eventually the assumption could be discarded and replaced with a classical physical explanation. He spent 8 years from 1900 attempting to accomplish that, without success, and finally, he had to accept that energy had an inherent 'graininess' totally unexpected and inexplicable with classical physics.

Experimental evidence had to wait for Einstein who, in 1905, submitted one of his 'miracle papers' to the "Annalen der Physik" that explained, with the help of Planck's quantum theory, the so-far mysterious phenomenon of the "photoelectric effect".

The effect was well known: ultraviolet light shined on a metal surface produces an electric current, liberating electrons from metals.

The mysterious nature was discovered by Philipp Lenard, a Hungarian physicist in 1902. He found that the number of electrons liberated from the metal in a given time, is proportional to the intensity of the illumination, but the energies of the individual electrons are dependent only on the frequency of the light.

Einstein's explanation: "According to the concept of the incident light consists of [photons] of magnitude of hv… one can conceive of the ejection of the electrons by light in the following way. [Photons] penetrate the surface layer of the body [metal] and their energy is transferred, at least in part, into kinetic energy of the electrons."

This explains why the number of electrons depends on the intensity of the light (the number of photons hitting the metal) but the energy of the individual electrons depends only on the energy of the individual photons, which is proportional to the frequency of the light, according to Planck ~hv.

As it turned out, Einstein received his only Nobel prize for this interpretation of the photoelectric effect, rather than his relativity theory. That was still too fantastic for the physics community.

Planck was at last vindicated as the "reluctant revolutionary" when he wrote: "A new scientific theory does not triumph by convincing its opponents and making them see the light, but rather because…a new generation has grown up that is familiar with it."

Niels Bohr (1885-1962)

Niels Bohr was born into a professional academic family and that's always a good start in life if you have the talent to take advantage of the opportunity. His father was a professor of physiology at the University of Copenhagen. Both Niels and his brother Harald received the best of education and they both excelled in their studies. Harald chose Mathematics, Niels chose Physics, but they shared one common passion: they both played football (called 'soccer' in North America). Harald was actually good enough to be in the Danish Olympic team that went on to win the silver medal in the 1908 summer Olympics. Niels was the reserve goalie and didn't quite make the team. In 1903 Bohr entered the University of Copenhagen and he did extremely well in his studies. In 1905 he entered a competition about the surface tension of liquids and shared the first prize gold medal, after much experimentation and theoretical work. His paper, titled "Determination of Surface Tension of Water by the Method of Jet Vibration" was printed in the 'Philosophical Transactions' of the Royal Society of London. For a 24-year-old Danish student that was a major achievement.

In 1897 J.J.Thomson had discovered the electron, as a tiny particle with a specific mass and charge which could have come only from inside the atom. This discovery shattered the prevailing idea of the atom as a single, unbreakable 'billiard ball'. Bohr was so fascinated by this revolutionary topic that he decided to write his Master's thesis on the various physical properties of metals, in the light of the new electron theory. He was so taken by the image of the atom that he decided to continue the same topic in his doctoral thesis and he successfully completed it by 1911. In between his Master's and doctoral dissertation, he met his future wife, Margarethe, who became his helpful assistance in all future publications.

Being a Doctor of Physics, he had only one path to follow: further his education abroad in one of the greatest centers of Physics at the time: The Cavendish laboratory in Cambridge, England, under the guidance of J.J. Thomson whom he greatly admired. They didn't get along very well because Bohr thought that the best way to impress J.J. was to point out some errors he had discovered in Thomson's work. That was a mistake that fully illuminated Bohr's character: he worshipped truth above all else and naively thought that all scientists were as open to learning from their own mistakes as he was.

His tenure with Thomson didn't last long. He was interested in Rutherford's discovery of the atomic nucleus and he asked to join the great New Zealander. Rutherford was duly impressed with the young Dane's credentials and intelligence, so he invited him to Manchester. Bohr arrived in April 1911. He was at the best place at the best of time. The experimental results recently obtained about the internal structure of the atom required a theoretical analysis and modeling of what the atom looked like inside, however, none existed yet. Niels Bohr jumped to the challenge to be the first one to provide it.

The newly discovered internal structure of the atom led Bohr to speculate about the possibility of explaining the differences between the chemical and physical properties of the elements. After all, elements were all made of atoms and their differences had to do something with the internal structure of the atoms. He suggested to Rutherford that these differences could be caused by the arrangement of the electrons around the nucleus. This was the first time when anyone suggested a model of the atom resembling a miniature solar system in which the electrons were freely rotating around the nucleus like tiny planets.

Bohr's idea wasn't unreasonable: the nucleus contained most of the mass of the atom, (just as the sun in our solar system) and the electrons were not embedded in the atom as was originally proposed in Thomson's 'plum pudding model'. By that time, it was known that the nucleus made up a very tiny central volume of the atom, so the electrons had to be on some outer shell to account for the size of the atom. The electrons had to be moving freely at some distance from the nucleus, otherwise, they would have fallen into the positively charged nucleus, destroying the atom in a catastrophic collapse. On the other hand, if the electrons were orbiting the nucleus, with enough speed, then the centrifugal force would keep them in their orbits, according to classical physics theory.

There was just one problem: according to electrodynamic theory, an orbiting negative charge would have to radiate electromagnetic radiation that would result in loss of kinetic energy of the electron and so it would have to spiral into the nucleus eventually, destroying the atom in the process. Of course, this doesn't happen, and this was a major stumbling block for Bohr's theory.

Bohr put this problem to the back of his mind and still thought that the differences between chemical elements (and their atoms) had to do something with the arrangement of the electrons inside the atoms. Rutherford who was a die-hard experimentalist and was suspicious of theoretical speculations without experimental data to back them up cautioned Bohr against unsupported theories.

As it turned out, Bohr was right, and in 1912 Frederick Soddy, Kasimir Fajans, and A.S.Russell provided the experimental proof. They found that when an element emits a beta particle, which is a high-energy electron, the element is transformed into the element with the chemical properties of the next element above it in the periodic table. The reverse was also true: when an element emits an alpha particle, which has two positive charges, it transforms into the element two places below it in the periodic table. That phenomenon became known as the "radioactive displacement" law. This was experimental proof that the physical and chemical properties of the elements were tied to the arrangement of positive and negative charges inside the atom.

Further experimental data suggested that if the mass of the atom was reduced without a change in the positive and negative charges inside it then there was no noticeable change in the element's physical and chemical properties. That made the task of separating some of the radioactive materials by chemical means, as George von Hevesy, one of Rutherford's Hungarian collaborators found out, nearly impossible. A new word "isotope" was coined to describe elements that had the same chemical and physical properties but different atomic weights. Soddy who wrote up the paper announcing this discovery received a Nobel prize in chemistry a few years later.

Bohr had the consolation of having been on the right track before Rutherford discouraged him. A good sport and a very nice person, he never resented this setback but proceeded to develop his atomic model based on the planetary arrangement idea. Bohr had to find some way to retain as much of the classical physics as possible and he needed help to solve the problem of collapsing atoms.

This help came from the first true 'epicyclist': Max Planck, who also had to abandon part of classical mechanics in order to develop his theory on black body radiation. Since Bohr saw no other way out of his problem, he started seriously considering Planck's quantum theory. As he wrote sometime later: "In the spring of 1912 I became convinced that the ... constitution of the Rutherford atom was governed throughout by the quantum of action."

At this point Bohr needed a little bit more than speculation, so he started on a series of detailed mathematical calculations to see if using Planck's theory could explain some of the experimental data, regarding the spectral lines of the hydrogen atom. His calculations needed some fundamental assumptions and he made two postulates.

Science has often proceeded by making assumptions and checking the mathematical and logical consequences. This approach was often used by the greats of classical physics, and it has proven useful over time. Maxwell's electrodynamics, for example, used it by introducing an extra term in his fourth equation that suggested the existence of electrodynamic waves. The entire body of relativity theory is based on Einstein's postulate of the universal speed of light. Mendeleev made up his periodic table without knowing if the predicted missing elements would ever be found (they were) and with the exact properties they were supposed to have (they do). These assumptions were based on *convincing* philosophical arguments and not made in order to assure that the predictions of the equations would match existing experimental data.

Both Planck and Bohr made their crucial equations by calculating backward from known data and adjusting their equations to assure that the predictions of their formulas would match what was already known. It's not exactly cheating, but not as pure science as when a set of compelling assumptions are made, without knowing what the consequences would turn out to be. Quoting from William Cropper's "Great Physicists":

"Bohr did not pick these equations out of theoretical thin air. He was guided by the observed patterns in the radiation spectra emitted by elemental substances, particularly atomic hydrogen."

So, after these caveats, let's see what assumptions Bohr based his calculations on:

Postulate #1: "An atomic system possesses a number of states in which no emission of energy takes place, even if the particles are in motion relative to each other, and such emission is to be expected in ordinary electrodynamics. The states are denoted as 'stationary' states of the system under consideration."

Postulate #2: "The dynamical equilibrium of the system in stationary states is governed by the ordinary laws of mechanics, while the laws do not hold for the transition from one state to the other."

Bohr still imagined the electrons, like little planets, orbiting the nucleus in circular or elliptical paths, according to Newton's classical mechanics but, at the same time, declared Maxwell's classical Electrodynamics invalid inside the atom. He maintained this contradiction in order to save his atomic hypothesis. Once he made this decision, he was free to proceed and describe what was taking place inside the atom.

He realized that it takes an amount of energy to liberate an electron from the atom (against the electrostatic attraction between the positive nucleus and negative electron), so he created the concept of the electron's *binding energy*. Then he went further: he declared that energy transfer takes place when the electron 'jumps' from one stationary state to another (discontinuously from one orbit to another) inside the atom. So far so good but then he was stuck. He needed another clue to proceed.

It came from one of Bohr's friends, H.M. Hansen, who asked Bohr if his idea of the atom's inner structure could account for the experimentally observed spectral lines of the atom. Bohr's original reaction was negative, he didn't immediately see how it could be possible. But then, Hansen explained the Balmer formula that Bohr was not familiar with.

Balmer, studying the distribution of the spectral lines of the hydrogen atom, realized that they were not random, but showed a mathematical regularity and managed to create a formula that precisely described their positions on the spectrum. Balmer realized that the frequencies of the spectral lines in the hydrogen atom could be described by the following formula:

$$v = R(1/2^2 - 1/n^2)$$

where R is a parameter determined from experiments and n is an integer, larger than 2: n=3,4,5....

Bohr suddenly realized that the differences between energy values of the hydrogen atom's stationary states were exactly described by Balmer's formula. This could not be a coincidence. As he said later: "As soon as I saw Balmer's formula, the whole thing was immediately clear to me."

He suggested that the spectrum's color band could be produced if an electron jumped from one stationary state to another, emitting a burst of energy that corresponded to the energy difference between the two states. From classical mechanics, he derived an equation connecting this binding energy with the frequency of rotation around the nucleus.

Bohr needed another clue so that he could derive the hydrogen atom's spectral line frequencies. This time it was Max Planck who came to Bohr's rescue. As Bohr wrote: "It seems necessary to introduce in the laws in question a quantity foreign to classical electrodynamics, i.e. Planck's constant…" He had to account for the discontinuous nature of the hydrogen atom's spectrum and thought that Planck's idea of energy packets (discontinuous radiation) just might do the trick.

This notion was also reinforced by the Balmer formula's discontinuous nature, which introduced the integer values of 'n' – referring to distinct electron orbits as Bohr saw it. So, combining the two equations (one classical, one quantum) he arrived at his formula for the radiation frequencies that described the hydrogen atom's spectrum with remarkable accuracy.

Bohr was aware of the nature of his proof for his formula, as he wrote: "The model has ben deduced, or if you prefer guessed, from experiments, not from theoretical foundations."

As William Cropper comments: "He could introduce an 'extramechanical' postulate that pictured electron jumping discontinuously from one stationary state to another but could not part with the classical picture of electrons in continuous orbital motion. What Bohr proposed was only half of a complete atomic theory – a theory that was only half crazy enough."

As Hendrik Kramers, one of Bohr's assistants commented: "The quantum theory has been very much like other victories; you smile for months, and then weep for years." Still, Bohr's atomic model and his derivation of the hydrogen atom's spectral lines was his crowning achievement and he deserves full credit for his creativity.

The rest of Bohr's life was a very active one indeed. He established a world-class educational institute in Copenhagen, *The Bohr Institute* where the best young talents in theoretical physics could continue their education and research. His colorful path included a rescue from Nazi-occupied Denmark in the fuselage of a British bomber plane that whisked him away to safety in England. He won the Nobel Prize (the gold medal had to be dissolved and stored in liquid form to hide it from the Nazis), advised President Roosevelt, Truman, and British PM Churchill, and warned the human community of the terrible dangers of nuclear arms race.

Quantum Mechanics

It's hard to determine who uttered the words "Shut up and calculate!" that so aptly describe Quantum Physic's future after Bohr's speculation about the inside of the atom as a miniature solar system.

Niels Bohr was the last of the Quantum Physicists who tried to visualize what's happening inside the atom. His model of a miniature solar system did not explain everything and did not please everyone. What followed was an abandonment of modeling attempts and was replaced by mathematical descriptions that could be used in calculations, in order to predict the behavior of very practical applications in industry and in weapons research. It was bound to happen, consistent with the centuries-long tradition of theory versus practice.

Just like Maxwell gave up all attempts to model the hypothetical aether as a contraption of springs and rollers and pullies and started from scratch by summarizing all that was known about Electrodynamics with his four equations – something that could be used for calculating practical problems, Werner Heisenberg and Erwin Schrodinger came up with complete mathematical treatment, seemingly completely different but, in effect, proven to be equivalent descriptions of quantum systems.

What's worse, probability entered physics in a big way for the first time. Physicists had to admit defeat in their attempt to describe exactly what was happening inside the atoms. Now they had to deal with probability amplitudes to determine what was likely to happen under certain circumstances. No wonder Einstein could never accept the new Quantum Mechanics as the final word on the subject.

The long line of exciting new experimental discoveries by Oersted, Faraday, Marie Curie, Rutherford, Chadwick, and Meitner came to an end with small advances here and there.

An excellent book about the rest of quantum physics is "Quantum Ten – A Story of Passion, Tragedy, Ambition, and Science" by Sheilla Jones. She described the two most prominent physicists in her introduction as follows:

"Erwin Schrodinger (1887 – 1961) An enthusiastic womanizer who was constantly getting distracted from his innovative work in science by the pleasures of the flesh."

Werner Heisenberg (1901 – 1976) - A handsome, ambitious young man whose self-confidence was continuously undermined by his father's anxious concern that he really wasn't smart enough to be a physicist."…

These two names are familiar even to the general public by two of their unforgettable contributions. Everyone has heard of Heisenberg's Uncertainty Principle that suggests that we will never get to the bottom of things; and Schrodinger's cat who is neither alive nor dead until we open the box.

Most textbooks on Physics history proceed with the great names in Quantum mechanics and their mathematical wizardry by which they produced two different, consistent, and, as it turned out, equivalent, descriptions of quantum systems.

Erwin Schrodinger (1887 – 1961)

The ultimate tool for the "shut up and calculate" generation of physicists and engineers is what is known today as 'Schrodinger's equation':

$$\nabla^2 \Psi + 2m/h^2 (E - V) \Psi = 0$$

This mysterious-looking equation opened the path to all our modern electronic technology giving us computers and the internet, as well as weapons research technology giving us nightmares. This was primarily Erwin Schrodinger's brainchild, built on Louis de Broglie's brilliant insight and contributed to by Paul Dirac and Max Born.

Louis de Broglie was from a French aristocratic family with a strong tradition of interest in science. Actually, his older brother, Maurice became an experienced experimental physicist, working on x-ray diffraction. This diffraction showed an interference pattern that was uniquely wave-characteristic, but it was also known, from the experiments of a British physicist, William Bragg, that x-ray radiation (under certain conditions) displayed particle-like characteristics. Maurice passed this seeming contradiction on to his brother in the early 1920s and de Broglie was instantly interested – here was a beautiful symmetry that appealed to him. If light radiation could be seen as particles (photons), as Einstein proved about the photoelectric effect, why couldn't particles, such as electrons, be seen as waves? He developed this theory in his doctoral dissertation in 1924. His professor couldn't make head or tail of it and sent it off to Einstein to solicit his opinion. Einstein wrote: "he lifted a corner of a great veil". Einstein's endorsement created great interest in de Broglie's thesis and greatly inspired Schrodinger to develop his own wave mechanics theory. Bohr and other physicists in his Copenhagen Institute initially dismissed the idea and the great feud between the Einstein and Bohr followers were set in motion.

De Broglie's ideas were confirmed in 1927 by Clinton Davisson and Lester Germer in the US, who were the first to experimentally observe electron waves. And, of course, it did not take long before electron microscopes were developed in the 1930s, using the electron's wave-like property.

Erwin Schrodinger was born in Vienna His father, who was the owner of a linoleum factory, had a keen interest in all sciences and arts and passed it on to his only son. Erwin started his formal education in the Akademische Gymnasium where he showed early talent in mathematics, physics, and languages. He became fluent in English, French, and Spanish. After high school, he enrolled in the University of Vienna to study theoretical physics. He became interested in atomic theory but was repelled by existing explanations that, as he later wrote: "its inherent contradictions sounded harsh and crude when compared with the pure and inexorably clear development of Boltzmann's reasoning." This quote shows his early interest in philosophical clarity and he did consider a career in philosophy for a while.

His first major academic appointment was, in 1921, at the University of Zurich where he stayed for six years and completed his wave mechanics theory. He started out with a clue from the "short but infinitely far-seeing" remark by Einstein about the dual particle-wave nature of light. `In Einstein's view matter (electrons and other particles) were "isolated fragments of energy" as proven by the well-known matter-energy equivalence formula from Einstein's special theory of relativity:

$$E=mc^2$$

On the other hand, Planck's quantum theory assigns discrete values to energy quanta according to his assumption of

$$E=hv$$

where E is the energy and 'v' is the frequency of the radiation. De Broglie combined these two equations as:

$$E=hv=mc^2$$
$$p=mc=hv/c$$

where 'p' is the photon's momentum p=mc (since 'c' is the speed of the photon) and converting frequency to wavelength according to the v=c/λ relationship, we get

$$p=h/λ$$

connecting the concept of momentum (particle attribute) and wavelength (wave attribute).

That's how far de Broglie got in establishing particle-wave duality that initially applied to photons. But there was no reason why this could not be expanded to other particles such as electrons, as well. And that is exactly what Schrodinger did.

He was using three other resources to build on: Hamilton's theory of light-ray and particle motion; the basic differential equation of optics - and Planck's energy-frequency equation. (William Hamilton was an Irish mathematician and physicist who worked out his theory almost a century before Schrodinger's time).

The mathematical derivation is beyond the scope of this narrative but, at the end, it produced the equation this chapter started out with. The equation looks like any other equation describing waves of any kind: water, sound, light (EM radiation). It also restates the conservation of energy principle by stating that the total energy of the system (like an atom) is the sum of the atom's kinetic and potential energies.

Solving the equation, we obtain the wave function:

$$Ψ(x,y,z,t)$$

This wave function describes the space and time-dependent property of the system (such as a hydrogen atom). The remaining question is: What does it mean? How do we interpret this wave function as applied to atoms and electrons? Schrodinger, then Born and Pauli interpreted it as follows: "For a single electron in a free atom, the wave function squared: $Ψ^2$ measures the probability of finding the electron at, or near, at the (x,y,z) location in a given time instant 't'.

He published 6 papers on the subject between 1925 and 1926. The first paper was written during his two-week Christmas vacation in the Swiss Alps. He was joined there by a mysterious woman, "an old girlfriend in Vienna" he said, whose identity is still a mystery, though she must have been a great inspiration because he finished his paper in two weeks.

He started his paper with: "In this communication, I wish first to show in the simplest case of the hydrogen atom that [Bohr's] rules for quantization can be replaced by another requirement, in which mention of 'whole numbers' no longer occurs. Instead, the integers occur in the same natural way as the integers specifying the number of nodes in a vibrating string. The new conception can be generalized, and I believe it touches the deepest meaning of the quantum rules."

When he applied his equation to an electron in the proximity of a proton attracting it by an electrostatic force, he could easily deduce Bohr's quantum rules for the hydrogen atom. Schrodinger never liked Bohr's arbitrary postulates that were not justified in any way other than making Bohr's equation match the observed result of the hydrogen spectrum. Now the same results were deduced from physical principles. As Sheldon Glashow writes: "Bohr's magic gave way to Schrodinger's method". He also confirmed Bohr's guess that the angular momentum of the electron is quantized.

Glashow goes on to write: "The Schrodinger equation applies to many other situations in physics. His second paper presented quantum-mechanical solutions to an assortment of classical problems, such as the vibrations and rotations of a molecule. In later papers, he explored the effects of electric fields on atomic spectra and examined various scattering problems. There seemed to be no limit to the versatility of his new equation and to the power of wave mechanics."

In the fall of 1926 Schrodinger had a telling conversation with Bohr as recounted by Heisenberg:

Schrodinger: "It is claimed that the electron in a stationary state of an atom first revolves in some sort of an orbit without radiating. There is no explanation given of why it should not radiate according to Maxwell's theory. Then the electron jumps from this orbit to another one and thereby radiates. … the whole idea of quantum jumps must simply be nonsense".

Bohr: "You are right in what you say but that does not prove that there are no quantum jumps. It only proves that we can't visualize them...."

Schrodinger: "If we are still going to have to put up with these damn quantum jumps, I am sorry that I ever had anything to do with quantum theory."

Schrodinger never fully accepted the quantum theory he helped to create. He wasn't alone in his doubts – Einstein was the best-known critic of quantum mechanics as the final word on the subject.

Isaac Asimov writes in his Biographical Encyclopedia of Science & Technology:

"...it occurred to him that the picture of the atom as built up by Bohr could be modified .. the Bohr atom might even be improved. In Schrodinger's atom, the electron can be in any orbit around which its matter waves can extend in an exact number of wavelengths. This produced a standing wave and therefore did not represent an electric charge in acceleration, so the electron, as long as it remained in its orbit, need not radiate light and did not violate the condition of Maxwell's equation. Furthermore, any orbit between two permissible orbits where a fractional number of wavelengths would be required is impossible. This accounts for the existence of discrete orbits with nothing possible in between, as a necessary consequence of the properties of the electron and not as a mere arbitrary deduction from spectral lines [as Bohr had done it]"

He received the Nobel prize in Physics in 1933, sharing it with Paul Dirac - "for the discovery of new productive forms of atomic theory."

The rest of Schrodinger's career wasn't as spectacular as this burst of creativity he displayed in Zurich. In 1927 Max Planck retired and asked Schrodinger to take over his post in Berlin. Schrodinger accepted and enjoyed his post until Hitler came to power. Schrodinger was profoundly anti-Nazi and saw the writing on the wall. He was one of the very few German scientists who, in spite of not being Jewish, decided to emigrate. He went to Oxford, Graz, and Rome before finally accepting directorship in the new Irish Institute for Advanced

Studies in Dublin. In 1956, with failing health (he had tuberculosis) he returned to Vienna where he died in 1961.

He left a widow Anny Bertel and a large number of grieving mistresses behind. All his life he was fascinated by women and sexual experiences that helped him achieve 'transcendence'. He and his wife had an unconventional marriage arrangement in which both were free to pursue extramarital affairs, and this did not affect their admiration and love for each other. Rules common for mortal people do not apply to geniuses, it appears.

Werner Heisenberg (1901 – 1976)

William H. Copper in his book "Great Physicists" starts the chapter on Heisenberg by saying:

"The birth of the grand synthesis of quantum theory – now known as quantum mechanics – was not the happy event it might have been. …Erwin Schrodinger and his colleagues in Munich and Berlin who claimed the child they called "wave mechanics" found little to admire in the other child, called "matrix mechanics" claimed by Werner Heisenberg and his friends in Gottingen and Copenhagen."

Sheilla Jones starts her book "Quantum Ten – A Story of Passion, Tragedy, Ambition, and Science" by bluntly stating that "Theoretical physics is in trouble". Then she goes on to say: "It may seem quite bizarre that modern physics would seek progress by regressing to a pre-Newtonian form of scientific methodology…The seeds of the shift currently taking place in science were sown eighty years ago from 1925 to 1927" She was writing about the fifth Solvay conference that was called in 1927 to reconcile two seemingly irreconcilable philosophies for understanding and describing nature and, within it, the quantum world.

Schrodinger's wave mechanics tried to visualize, at least to some extent, what was happening inside the atom, while Heisenberg gave up all attempts to see what was actually going on and treated atoms as tiny black boxes that no one could peer inside. What mattered for him was representing atoms with abstract mathematical symbolism that could be used for calculating probable outcomes of experiments.

He started out w th the experimental data of spectral lines and tried to devise a mathematical representation that would account for them. Following Bohr's assumptions, he connected the spectral lines' frequencies with the bincing energy of the electrons. So, when an electron 'jumps' from a higher energy, designated by E_2 to a lower energy E_1, the frequency of the spectral line created will be, following Planck,

$$V_{21} = (E_2 - E_1) / h$$

Then he came up with a new concept: introducing statistical probability into the picture by assuming a hypothetical number A_{21} that would be the probability of an electron jumping from the E_2 energy level to the E_1 level. This was a reasonable assumption, but there was no way to estimate or measure this number. However, Heisenberg wanted to explore this line of reasoning and see where it would lead him. He generalized this for an atom with three energy states and arranged the hypothetical probabilities in a square array as follows:

$$A_{11} \quad A_{12} \quad A_{13}$$
$$A_{21} \quad A_{22} \quad A_{23}$$
$$A_{31} \quad A_{32} \quad A_{33}$$

where row 1 corresponded to the probabilities of electrons jumping from state E_1; row 2 was for E_2 and row 3 was for E_3. Then he started manipulating this 'matrix' with 'Fourier analysis', well known from mathematics, to see where it would take him. So far this was a purely symbolic exercise without any physical content. His aim was to deduce the Balmer formula for the Hydrogen atom spectrum from his mathematical manipulat ons.

However, he ran into a problem because his math would only work if he assumed some very strange and unusual algebraic rule: the multiplication he had to perform between two matrices could not be commutative. This means that two matrices x and y had to be multiplied in a certain order, meaning that xy wasn't equal to yx. This stomped him. He still finished his theory with this strange multiplication rule built-in, but he felt very uneasy about 'cheating', as far as he could tell, in his calculation. Another example of epicycle physics, but he was in luck.

Max Born, who was originally a mathematician before becoming a physicist knew matrix algebra. Teaming up with another mathematical vizard, Pascual Jordan, the two of them published a separate paper just on the mathematical proof of Heisenberg's system in 1925. As it turned out, to Born's great consternation, another paper was published, only 9 days later, in an English journal, titled "The Fundamental Equations of Quantum Mechanics". It was written by an English physicist, Paul Dirac, and used the exact same non-commutative matrix algebra that Born and Jordan had worked out.

When Heisenberg found out about this unexpected development, he sent a letter to Dirac: "Now I hope you are not disturbed by the fact that indeed parts of your results have already been found here some time ago and are published independently here in two papers – one by Born and Jordan, the other by Born and me…On the other hand, your paper is also written really better and more concisely than our formulation given here."

Dirac wasn't disturbed but went ahead and used the new matrix mechanics system to deduce the Balmer formula for the hydrogen atom. He succeeded and published his result, as the final proof that Heisenberg's model was a real quantum mechanical system. The great advantage of Heisenberg's system over Bohr's atom was the total lack of invisible orbits (replaced by 'states') or mysterious jumps from orbit to orbit (replaced by 'transitions' from one state to another). The disadvantage was the total lack of physical meaning and visualizability of what was happening inside the atom.

So, in a short span of time, two completely different systems were proposed to explain the behavior of quantum systems. As it turned out later, they were stating the same relationships, equivalently. Both could be used in calculations (Schrodinger's was easier to use) and both produced the same results.

The authors both hated the other's system. Schrodinger declared: "I was discouraged, if not repelled, by what seemed to me a rather difficult method of transcendental algebra, defying any visualization." Heisenberg wrote in a letter to Pauli: "The more I think about the physical portion of the Schrodinger theory, the more repulsive I find it." The battle line has been drawn between the two camps: Heisenberg and his friends in Gottingen and Copenhagen, versus Schrodinger and his colleagues in Munich and Berlin.

Heisenberg was 24 years old when he created one of the two new quantum mechanical systems and this catapulted him into the ranks of world-class physicists. He received the Nobel prize in 1933 "for the creation of quantum mechanics, the application of which has, inter alia, led to the discovery of the allotropic forms of hydrogen."

Heisenberg continued his mathematical treatment of quantum systems and discovered another great law that he called "The uncertainty principle" and he published it in a paper in 1927. The principle basically states that it is impossible to make an exact measurement, at the same time, of certain dynamically linked values such as position and momentum or energy and time. He even provided a formula for the measure of this imprecision:

$$\Delta p_x \, \Delta x >= h/2\pi$$

for momentum and position and

$$\Delta E \, \Delta t >= h/2\pi$$

for energy and time.

Two different interpretations have been attached to the uncertainty principle. The objective and pragmatic interpretation recognize that it is impossible to measure quantum particles (such as electrons) with any precision because any measuring process humanly possible necessarily disturbs the state of the measured particle, so if we measure momentum, then the following measurement of the position will be unavoidably inaccurate.

The philosophical interpretation argues that this principle abolishes the millennia-long assumption of cause and effect chains and makes the world nondeterministic and random. Personally, I shy away from philosophical interpretations and stick to what we actually know and that is summarized by the formulas above that Heisenberg was able to deduce from his quantum mechanics.

Heisenberg was appointed to the position of professor of theoretical physics in Leipzig, in 1927, where he continued his research, even during the Nazi era. Following Planck's advice:

"You cannot stop the catastrophe, and in order to survive, you will be forced to make compromise after compromise. But you can try to band together with others and form islands of constancy. You can gather young people around you, teach them to become good scientists, and thus help preserve the old values…for such groups can constitute so many seed crystals from which new forms of life can grow."

Planck for sure followed his own advice and so did Heisenberg. They both stayed in Hitler's Germany, trying their best to keep their values. In 1939 Heisenberg was ordered to join the Uranium Project, to explore the possibilities presented by Otto Hahn's fission experiment. They were by then cut off from the world's most famous and productive scientists and could not make much progress in a scientific vacuum.

Heisenberg survived the war and became the director of the Max Planck Institute in Gottingen. For the rest of his career, he pursued every physicist's holy grail: the unified field theory, but in this, he failed like every other physicist (including Einstein) who were seduced by the dream of one equation that would explain everything.

Richard Feynman (1918 – 1988)

In Book I (*Humane Physics – Classical Physics*) I commented on Feynman as: "Richard Feynman, the bongo-drum-playing prankster with the charm of a curious child, the soul of a poet and a mind like a laser beam". I still stand by that description. Several biographies, movies, quotes, and hundreds of stories circulate about this most colorful genius. He published two collections of anecdotes about himself (*Surely You're Joking, Mr. Feynman!: Adventures of a Curious Character* and *What Do You Care What Other People Think?:*

Further Adventures of a Curious Character) and his 3-volume "*Feynman Lectures on Physics*" are a world-famous resource for students everywhere. He ventured beyond Physics with his little-known volume "*The Meaning of It All: Thoughts of a Citizen-Scientist*" in which he gives the world common-sense (therefore hopeless) solutions to all of humanity's 'unsolvable' problems.

　　Feynman is credited with being the most creative and original of all physicists, at least in the 20th Century. He had a unique way of looking at a problem and coming up with the most unexpected and unusual solution. That trait manifested in him very early on, even in high school which he found very boring, so he spent his class time self-educating in advanced calculus from books his teacher gave him to keep him from making smart-ass comments, disrupting the lecture.

　　The first thing in physics that intrigued him was the *principle of least action*. It is a very seldom touched-on topic in Newtonian mechanics that states that the actual path followed by a moving object corresponds to the minimum value of the difference between the kinetic and potential energies integrated over all the actual paths the object could take. This is what is called 'action' in Mechanics. Feynman used the concept of action in all areas in physics (including quantum mechanics) to solve seemingly 'unsolvable' problems. As he wrote: "Instead of differential equations, it tells the property of the whole path. And this fascinated me. That was the greatest thing ever. The rest of my life I have played with action, one way or another, in all my work. I loved it always."

　　After high school, Feynman considered his options. First, he thought of becoming a mathematician but, when he asked what he could do with a math degree other than teaching it, he was advised that he could do actuarial calculations for life insurance companies. This did not inspire him, so he settled on physics instead.

　　He did his undergraduate studies at MIT and then followed with his master's at Princeton. He was offered a research assistantship with John Archibald Wheeler who was only six years older than Feynman, and the two of them got along very well. He again used the concept of action there to calculate the probability of an electron passing in space from one point to another by summing up (integrating) contributions from every conceivable path the electron could take. Wheeler was so impressed that he brought it to Einstein's attention who commented: "I still cannot believe that God plays dice, but maybe I have earned the right to make mistakes".

In 1942 Feynman joined the Manhattan Project and became the youngest group leader to solve difficult computational problems. In addition to his contribution to the atom bomb project, he spent his Los Alamos years driving the security people crazy, time and time again, by breaking into top-secret safes, just to prove to himself that he could. His first wife, Arline, slowly dying of tuberculosis in a nearby sanatorium, occupied the rest of his time, every moment he could spare from his work, till the very end of her life. She was his childhood sweetheart and the one true love of his life.

After the war Feynman spent five years at Cornell University, followed by his final move to California where he became a professor of theoretical physics at Caltech. One accomplishment he is not very well known for was his contribution to the solution of the Space Shuttle Challenger's tragic end. It was Feynman who provided the critical clue that explained the reason for the explosion: the shuttle was launched in unusually cold weather (against the strenuous objections of the engineers) that affected the O-rings between the sections of the solid-fuel rockets.

It's hard for me to be objective about him because he has been my hero ever since my university years. Whenever none of my professors could explain clearly enough any particular mystery bothering me about some aspect of my studies, I could always count on Feynman to clarify things for me.

His most famous contribution to science is QED, or Quantum Electrodynamics. This is his greatest achievement for which he received the Nobel Prize in 1965. QED is essentially a calculation method for the probability of electron and photon interactions. No new particles were discovered, no new laws formulated, but it gave physicists and engineers a priceless tool to conduct their calculations and make their predictions.

At this point, I have a problem with my narrative. Do I go into a detailed description of this tool and probably lose most of my readers, or give them an idea of the significance of quantum mechanical calculations? What is it good for? How did it change the world? What intriguing philosophical insight have we obtained?

The answer to these questions (except for the philosophical insight one) is, in two words, *computers* and the *internet*. Including the one, I am writing this story on. Computers and quantum mechanics are very closely intertwined from the moment semiconductor-based computers appeared on the scene.

In 1931: "The Theory of Electronic Semi-Conductors" is Published by Alan Wilson, using quantum mechanics to explain basic semiconductor properties. Walter Schottky developed the first semiconductor component: the 'Schottky diode', which was followed by transistors and then integrated circuits (chips) that all computers today are based on.

William Bradford Shockley Jr. was an American physicist and inventor. Shockley was the manager of a research group at Bell Labs that included John Bardeen and Walter Brattain. The three scientists developed a new branch of quantum mechanics known as surface physics. they were jointly awarded the 1956 Nobel Prize in Physics for "their researches on semiconductors and their discovery of the transistor effect".

The newest advance in computer technology these days is the much-touted 'quantum computers' research using the mysterious and mystifying quantum mechanical effect known as 'entanglement' what Einstein called "spooky action at a distance". But that topic doesn't belong in the chapter on Richard Feynman and I won't go into it just yet. Today quantum mechanics is used in all areas of industry, both beneficial and harmful (weapons research). The "shut up and calculate" edict is still the primary directive for scientists and engineers applying quantum physics in all areas of development.

So, the question now is: Do we understand QED?

Feynman gave the most eloquent answer to that question in his book: *QED – The Strange Theory of Light and Matter*. I'll quote this answer from this book because it explains brilliantly what we mean by *understanding* physics:

"Everybody who comes to a scientific lecture knows they are not going to understand it...and you think I am going to explain it to you so you can understand it...No, you are not going to be able to understand it…You see, my physics students don't understand it either. That is because I don't understand it. Nobody does. …physicists learned to realize that whether they like a theory or they don't like a theory is *not* the essential question. Rather, it is whether or not the theory gives predictions that agree with experiment. …The theory of quantum electrodynamics describes Nature as absurd from the point of view of common sense. And it agrees fully with experiment. So, I hope you can accept Nature as She is – absurd."

Murray Gell-Mann – 1929 - 2019

I was hesitating whether to write this chapter at all because the topic of particle physics is, for me, boring as hell to say the least. A delightful story, told by William Cropper in his book *Great Physicists* recounts an exchange between Enrico Fermi and one of his students, asking him to name the fundamental particles. Fermi's response: "Young man, if I could remember the names of these particles, I would have become a botanist."

Originally there were electrons and then protons (named by Rutherford for the Hydrogen nucleus) and then neutrons, discovered by Chadwick in 1932 and the list seemed to end. Then all hell broke loose. Atom smashers came into vogue in a big way and, once these three basic atomic constituents collided at high speed, a whole new zoo of created particles started to emerge. Nobody knew if any of them were elementary particles or only by-products of these high-energy collisions. By the 1960-s there were nearly a hundred and physicists despaired trying to categorize them into a meaningful system.

Even the category names they came up with were hard to remember. Every particle, or system of particles, is either a *fermion* or a *boson*. Protons, neutrons, and electrons are fermion, while bosons may be either elementary, like photons, or composite, like mesons. Hadrons are any strongly interacting particle, either a fermion or boson. Then there are the baryons, mesons, leptons, pions, kaons, hyperons and the list goes on and on. And, of course, every particle ever discovered has its anti-particle as well. Dear reader, are you confused already, or maybe bored to the point of giving up? If so, I can sympathize. That's where I gave up on particle physics in my own education. The person, who did not give up on this confusion was Murray Gell-Mann, an American physicist who was fully determined to find some underlying order in this mad zoo of a confusing array of particles.

Gell-Mann was a prodigy as a child, he could multiply large numbers in his head at the age of three, won a spelling bee against 12-year-olds at the age of seven and by eight he was smarter than most college students. At fourteen he was accepted to Yale with a full scholarship where he decided to study physics because his first choice, archeology, was discouraged by his father and Murray wasn't interested in engineering. He went on graduate studies at MIT after he was not accepted at Yale. He later wrote that "A little reflection convinced me that I could try MIT and then commit suicide later if I wanted to, but not the other way around". He received his Ph.D. in 1951, at the tender age of 21.

His first job saw him at the Princeton Institute for Advanced Studies, working with Oppenheimer as an apprentice. As no faculty position was available, he moved over to Enrico Fermi's Institute of Technology (later called Caltech) and started his career in earnest. His main area of research was particle physics and he set out to find some order in the teeming crowd of elementary particles.

In 1961 he devised a mathematical theory he called the *eightfold way* because of the octets of particles in the classification (a reference to the Eightfold Path of Buddhism) based on symmetry arguments. This scheme found a place for all known particles, but with gaps left open for future discoveries, not unlike Mendeleev's periodic table. Several of the predicted particles were soon found and Gell-Mann has established his reputation.

In 1963, Gell-Mann proposed a specific model for the hadron's internal constituents, which he called *quarks* (a reference to the novel Finnegans Wake, by James Joyce) – the first time anyone hinted at a possible internal structure of protons and neutrons.

In 1969 he received the Nobel prize (not for his quark theory that was still too controversial) but "for his contribution and discoveries, concerning the classification of elementary particles and their interactions."

In the late 1960s, at the Stanford Linear Accelerator, protons were bombarded with high-energy electrons. The experiments showed hard, point-like charged particles within the protons. Feynman jokingly proposed the name *partons* but Gell-Mann would have none of it – they were his long-sought-after quarks.

Apart from physics, Gell-Mann had a wide range of interests in archaeology, birdwatching, and linguistics. In 1995 he wrote a popular science book about physics and complexity science, *The Quark and the Jaguar: Adventures in the Simple and the Complex* The title of the book is taken from a line of a poem by Arthur Sze: "The world of the quark has everything to do with a jaguar circling in the night"

Based on Gell-Mann's research, and contributed to by many other physicists, now we have what is called the *Standard Model of particle physics* that fully (more or less) explains the taxonomy and characteristics of sub-atomic particles. However, details of this model are way beyond the scope of this narrative, so I have to finish this part of the story here.

String Theory

String theory started with a young physicist, Gabriele Veneziano, at CERN (European Organization for Nuclear Research in Switzerland) in 1968. He was studying data from high-energy atom smashers and noticed a curious pattern. The data seemed to match the mathematical formula discovered by Leonhard Euler (Swiss mathematician). Veneziano wondered whether this was a coincidence or something deeper. He couldn't explain it, but two years later, Leonard Susskind of Stanford University made a startling suggestion: If the strong force binding the protons together inside a nucleus was due to a tiny, extremely thin strand that connects them, then the observed pattern in the data would follow the Euler formula. These tiny, elastic, rubber-band-like strands would be vibrating at different frequencies and they were envisioned as one-dimensional energy-packets, one hundred billion billionth the size of the atomic nucleus

(10^{-33} centimeters). This, at this point, seemed like pure speculation without any experimental confirmation and, what's even worse, even the possibility of evidence emerging later. The tiny Planck-length size of the proposed strings guaranteed that they would remain undetectable with any present or future technology. However, epicycle-physicists are not deterred with such consideration; they forged bravely ahead. Some of their calculations suggested that a new particle with mass zero and spin two would be produced in high-energy collisions (none were ever found) and that particle could be the theoretically predicted *graviton*, the transmitter of the gravitational force which is also supposed to be massless and have spin two.

This was the first-ever suggestion that quantum mechanics and gravitational force could be connected by a new theory. A very thin argument to support such a lofty connection. The 1974 paper submitted by John Schwarz, an American theoretical physicist, was ignored by the scientific community. Schwarz didn't give up but teamed up with Michael Green of Queen Mary College in London and they refined their calculations so as to resolve some earlier anomalies and contradictions present in their theory. By the summer of 1984, they had their revised theory and this time physicists were ready to listen. Physics was at an impasse, unable to make any further progress in unifying quantum mechanics and Einstein's general theory of relativity; the proposed string theory was the only one that offered a glimmer of hope, however tenuous, for making progress.

According to the theory, the properties of the large number of discovered particles could also be explained if it is assumed that *the only elementary particle* is the string itself and its different vibration patterns manifest as an electron, a quark, a neutrino, etc. How about reducing the zoo of a hundred observed particles to one simple central cause: vibrating strings and, to boot, explain the gravitational force itself?

The idea was so seductive that, almost overnight, thousands of physicists jumped on the bandwagon and started developing it further. However, the theory would only work if the physicists assumed that the world had *twenty-five dimensions of space*. The number was reduced to *only* nine later. Then a new particle, called a *tachyon,* had to exist that could travel faster than light and, finally, there had to be *massless particles* in the universe.

I have a problem with explaining string theory to non-physicists, because it is totally non-intuitive and, frankly, it uses too many wild assumptions. What does it mean in laymen's terms that space has 25 (or even 9) dimensions? We know 3-dimensional space – we have lived in it all our lives. We can even understand space-time as a four-dimensional arena for the events we observe, but 25 (or 9) dimensions mean nothing to me or my readers. I can accept that physicists had to make these assumptions in order that the mathematical deductions would result in the desired configurations. But how much can we trust this sleight-of-hand science that smacks of cheating? How many epicycles did Ptolemy need in order to explain the movement of the planets? Thirty-four altogether if I remember well.

Most popular books on Physics use colorful analogies and suggestive diagrams to 'explain' inexplicable concepts (like elastic rubber bands), but I throw my lot in with Richard Feynman when he wrote: "No, you are not going to be able to understand it…You see, my physics students don't understand it either. That is because I don't understand it. Nobody does..the essential question. … is whether or not the theory gives predictions that agree with experiment." As far as string theory is concerned, such predictions and experiments are hopeless now and foreseeable future.

However, some physicists are so enamored by the 'beauty' of their mathematics that they find it, by itself, the proof they need to validate their theory. Remember Clerk Maxwell who, without a shred of experimental data, constructed an elaborate mechanical clock-work model with invisible spinning cells and rotating, moving wheels that permeated all space? Yet, when he developed his mathematical equations, they described the magnetic Coulomb Law, the magnetic induction effect of an electric current, and the induction of an electric current by a changing magnetic flux. He made sure it was understood that his model was arbitrary, a result of his imagination, and not necessarily describing nature. I often wish that modern physicists had the same honesty and commitment to truth.

Cosmology - Epicycle Physics

As I stated in the introduction: "The third phase [of physics history] is pure speculation, based on wild assumptions that could never be verified. Very sophisticated mathematical models were created, the predictions of which could never be tested, and no conceivable application could be seen forthcoming. However, some Physicists see the proof of their theories in the aesthetic beauty of their equations. A word of caution: many theories in science, like Ptolemy's epicycles, were considered beautiful and also worked for a thousand years in predicting the planets' movements, in spite of it being utter nonsense. That is why I gave the title "Epicycle Physics" to this third phase of Physics".

Let's assume, for the moment, that multiverses are a science fiction fantasy or the product of some physicists' minds with colorful imagination, and let's define the word 'universe', as it is defined by the Cambridge English Dictionary as: "everything that exists, especially all physical matter, including all the stars, planets...."

So now we are looking at it from Earth as we have been doing it from the dawn of history and speculating about the pinpricks of light in the night sky. Actually, we are doing a lot more than that with our instruments both on Earth and in space. The earliest speculations were based on the assumption that it was all about us, that all those things in heaven were revolving around us in perfect circles, because 'God' was perfect and allowed only perfect circles (or spheres if you listen to Aristotle).

Then came Ptolemy who invented even more circles (he called them 'epicycles') to explain the irregular motions of the planets, inventing circle after circle that created the great Ferris wheel in the sky with circles around circles to account for the observations. See detailed explanation of Ptolemy's system in *Book I of Humane Physics.* Then real astronomy started with Galileo and science never looked back

What I often find amusing is how earnestly and seriously we are talking about stars and galaxies hundreds, thousand, millions of light-years away, without any assurance that those celestial objects are still there. What we see now are stars as they were hundreds, thousands, and millions of years ago, due to the time light requires to reach us.

However, science needs to make assumptions and one of the first assumptions of cosmology is that what we see today still exists out there. In *Book I – Classical Physics* I described the course of Astronomy leading from ancient Greece to Newton. Now it's time to pick up the narrative again, beyond Newtonian astronomy. The person most closely associated with modern astronomy is Edwin Hubble.

Edwin Hubble (1889 – 1953)

Hubble was a very ambitious man and he pursued his ambitions with great determination. He was a good student in school but not a prodigy - very good in sports and he obtained a scholarship to the University of Chicago in 1906. He wanted to study astronomy but his father, a dominating figure in his life, had other plans: he wanted Edwin to study law so he could make a good living. Hubble could not oppose his father's wishes, but he did not give up his ambitions for astronomy either, so he studied both science and subjects required to enter law school. Two of his teachers in Chicago were Nobel laureates Albert Michelson (who made the crucial experiment disproving the aether hypothesis) and Robert Millikan (who measured the electron's charge with the oil-drop experiment).

That stroke of luck prepared him for astronomy, but he also studied hard to qualify for a Rhodes scholarship, and in that he was successful. In 1910 he left for Oxford, England. Once there, he tried very hard to emulate anything and everything British. He changed his American accent, adopted what he considered British mannerism, smoked a pipe, joined a rowing club, and dressed accordingly to his newfound persona. He kept his word to his father and studied law, until 1913 when his father died. That obstacle removed, Hubble could pursue his lifelong dream of astronomy. He returned to Chicago following his father's death.

In 1915 he was granted "seeing time" in the University of Chicago's Yerkes telescope and he could start observing what most interested him: dim objects called nebulae. George Hale, former director at Yerkes, became director of the new Mount Wilson Observatory in California and offered Hubble a position there after he finished his dissertation in Chicago. However, after he successfully passed the oral examination, patriotic fever caught up with him and he volunteered for military service in 1917. His service in Europe did not last long but he saw some action and was even slightly wounded by an exploding shell.

Hale, expecting the 100-inch telescope very soon and needing all the help he could get, urged him to return to Pasadena. Hubble took up his position in 1919. The 100-inch new telescope was very high in demand by Hale's department, so Hale established a ritual where the astronomers, at dinner time, were all seated according to their daily assignments: at the head of the table sat the one assigned to the 100-inch scope (apparently a major honor), then farther down the table the one assigned to the 64-inch scope, etc.

Hubble's first quest was to measure the distance to a nebula in the Andromeda constellation. Measuring stellar distances was not an easy task at the beginning of the 20th century. The first successful measurement of a star's distance from Earth was done by Friedrich Bessel who measured the slight change of angle of observation of the star called 61 Sygni, six months apart (from opposite points of the Earth's orbit around the Sun and used basic trigonometry to calculate its distance to be around 100 trillion km.

Distances within the solar system were long-established by Hubble's time and even the diameter and thickness of the Milky Way galaxy were known approximately, but no one knew whether any stellar object could be found outside our own galaxy.

Nebulae (meaning "cloud" in Latin), have been known for centuries, were faint smudges without any star visible inside them, so there was no solid object that astronomers could focus on to try to measure the distance. John Herschel, the son of William Herschel, discoverer of Uranus, was convinced that nebulae were young stars inside the Milky Way, surrounded by debris that obscured and diffused their light. Opposing his view, Immanuel Kant, the German philosopher, believed that they were groups of stars far outside the Milky Way galaxy and the stars inside the nebulae were invisible due to the extremely vast distance from us. However, by Hubble's time, the 100-inch telescope revealed a large number of stars inside several nebulae.

In 1920 The National Academy of Sciences in Washington hosted a major conference to resolve the issue, once and for all, whether anything existed outside the Milky Way. This conference became known as the "Great Debate". Hubble's observatory was represented by Harlow Shapely, Hubble's arch-enemy, who claimed that the Milky Way was all that existed in the universe, while Heber Curtis of the Lick Observatory advocated for the opposite view. There were arguments aplenty on both sides, what they both lacked was experimental evidence. That's where Hubble came in. Secretly convinced that Shapely was wrong he set out to prove it. The clue was to be found in his main interest: variable stars that change brightness periodically. Some of these stars turned out to be binary stars that circle and eclipse each other, while others had a different distinctive pattern of their period: it was less symmetrical, and the graphs of their brightness had a different shape from the binary stars. One of these stars was called 'Delta Cephei', so the groups of variable stars similar to it were called "Cepheid Stars".

Studying Cepheid variable stars got a major boost by applying the newly discovered technique of photography which replaced the inaccurate hand drawings astronomers had had to rely on before. Now images taken by the telescope could be transferred to glass plates that could be studied, measured, and compared to each other.

Edward Pickering, director of the Harvard College Observatory was among the first to use the new technique. Over a decade he collected half a million photographic plates that all needed analysis. He hired a team of men, called 'computers', but found them too impatient to do the work properly. He exclaimed one day: "my Scotch maid could do a better job". So, he fired them all and replaced the whole team with young women. True to his word, he put his maid in charge of the new group, which was nicknamed "Pickering's Harem". These women proved much more reliable than the men and much cheaper to boot. Not only that, but they used their brains and noticed patterns that evaded even the highfalutin astronomers. Two of these 'computer' women, in particular, stand out with their discoveries.

Annie Jump Cannon cataloged about 5000 stars per month during four years, noting their location, brightness, and color and classified them in a system, still used today, the seven classes of O,B,A,F,G,K,M which students memorize by the mnemonic "Oh Be a Fine Guy – Kiss Me!"). In 1925 she received an honorary doctorate from Oxford University.

The other major contributor in "Pickering's harem" was Henrietta Leavitt, a deaf graduate of Harvard University's Radcliff College for Women. She volunteered for the job of unpaid 'computer' and was assigned the group of variable stars for cataloging. During her time at Harvard, she cataloged over 2400 variable stars.

She wasn't satisfied just by cataloging these stars; she looked for a pattern by plotting a graph of the apparent brightness against the period of variation for 25 Cepheid stars. She was astonished by the result: Cepheids with longer periods in their variation usually turned out brighter. In 1912 she announced this result as: "A straight line can be readily drawn among each of the two series of points corresponding to maxima and minima, thus showing that there is a simple relation between the brightness of the variables and their periods."

This discovery enabled astronomers to work out the *relative distances* of two stars because of a principle, long-established in optics, that brightness fades as the square of the distance. To determine the absolute distance from Leavitt's formula, all they needed to do was to measure the actual distance of *one* Cepheid star. Harlow Shapley accomplished this by using parallax and with this, astronomers had a way to measure the absolute distances of stars in the Milky Way galaxy. Leavitt's contribution was a major breakthrough in astronomy and, as I recounted in the chapter on Lise Meitner, she was seriously considered for the Nobel Prize in 1925, only to discover that she had already died three years earlier in obscurity.

On October 4 1923 Hubble was observing on the 100-inch telescope on Mount Wilson and found a Cepheid variable star in the Andromeda Nebula. This enabled Hubble to calculate, using Leavitt's and Shapley's formula, the distance of the star and therefore the nebula itself. The result of his calculation was roughly 900,000 light-years, far outside the known diameter of the Milky Way galaxy. And, if the nebula was that far away and still visible to the naked eye, it had to be so full of stars ("My God, it's full of stars" quoted from the movie *2001*: A Space Odyssey) that it only could be another "island universe" – another galaxy.

After successfully calculating the distance of another, dimmer, Cepheid in the same Nebula which confirmed his earlier result, Hubble was confident enough to announce his finding in a letter to Shapley (to rub it in, for sure) and Shapley commented: "Here is a letter that has destroyed my universe." In 1924 Hubble received a $1000 prize from the American Association for the Advancement of Science as the most exceptional paper of the year.

The universe already considered to be large already with an estimated diameter of the Milky Way galaxy suddenly exploded to a previously unimaginable dimension. However, Hubble wasn't finished yet with his astonishing results. More was to come. He used spectroscopy to analyze the spectrum of the stars. That wasn't new by itself because physicists had been using star spectra to determine the elements stars were made of – The first evidence of helium in the sun was observed on August 18, 1868 (way before the element was found on Earth) by spectral analysis.

The Doppler effect, announced by American physicist Christian Doppler in 1842 established a mathematical formula for the waves emitted by a moving object relative to an observer as

$$\Lambda_r = \Lambda \times (1 - v_e/v_w)$$

where Λ_r is the received wavelength by the observer; Λ is the wavelength of the emitted wave by the moving source; v_e is the speed of the emitter and v_w is the speed of the wave. In 1868, using the spectrum of the star Sirius, William and Margaret Huggins noticed that the spectral lines were shifted a tiny amount: each line represented a wavelength 0.015% larger than what had been observed about the sun. Conclusion: Sirius was traveling away from Earth.

An amusing story in Simon Singh's excellent book: *Big Bang* tells us a story about an astronomer trying to convince the police that he had jumped a red light because he was moving toward the light and it was blue-shifted and appeared to him green. According to the tale, the policeman then doubled his fine for speeding because, in order for the claim to be believable, he would have had to drive at 200,000,000 km/h. This must have been an extremely well-educated policeman!

Soon results started trickling in about Doppler shifts about nebulae. By 1917 Vesto Slipher, an amateur astronomer, calculated the relative speed of a dozen nebulae with most of them receding from Earth, and then 25 more were found all receding. When Hubble heard about these results, he set to work in earnest.

His assistant, Milton Humason had worked as a bellboy at the Mount Wilson hotel, later as a mule driver to carry supplies, then a janitor at the observatory. He was curious, so he educated himself in mathematics and astronomy, and in three years he was appointed to the photographic division. Two years later he became a recognized astronomy assistant, helping Hubble to analyze forty-six galaxies. They plotted the calculated velocities and the distances for each galaxy on a chart and Hubble noticed the pattern - the velocity of the galaxy was proportional to its distance from us and all of them were receding. In other words, the farther a galaxy was, the faster it sped away. In 1929 he published a paper titled "A Relation Between the Distance and Radial Velocity Among Extragalactic Nebulae."

He refused to speculate about an interpretation that already surfaced about a possible great explosion that would explain how and why the galaxies were receding from us; it postulated a possible origin of the universe at which time the galaxies were all close together and started their mad rush out to empty space.

In 1931 Hubble published a paper announcing the result of more refined observations of galaxies twenty times as distant as the ones he reported on in his 1929 paper. This led to the famous Hubble's Law as known today

$$v = H_0 \times d$$

where v is the velocity of the galaxy, H_0 is what is called now Hubble's constant calculated to be 558 km/sec/Mpc (Mpc for Megaparsec). So not only the galaxies are moving away from us but the speed, according to Hubble's law, is increasing. Something is pushing the observed universe apart at an accelerating rate.

Hubble resolutely (and wisely, in my opinion) stayed away from speculation about the 'Big Bang' theory. As he wrote: "The writers are constrained to describe the apparent velocity-displacements without venturing on the interpretation and its cosmological significance."

Hubble's discovery led to an example of what I am calling 'epicycle physics'. When Einstein arrived at his field equation for the General Theory of Relativity, in 1915, the equations unexpectedly predicted that the universe would collapse under its own 'weight', that is the gravitational attraction of all the stars and galaxies in it. Einstein didn't like this because it was in contradiction with the prevailing assumption that the universe was static and stable. So, he introduced an arbitrary term, now called the "cosmological constant" to fudge the equation to bring it back in line with the static universe model. However, after he visited Hubble at Mount Wilson in 1931 and had a chance to review the evidence of the expanding universe, he publicly renounced his own static universe model, abandoned the cosmological constant, and returned his equation to its original form. He considered the invention of this arbitrary term the "greatest blunder of my life".

Then comes the twist. Modern Big Bang theorists are eager to put the constant back in the General Relativity field equations because it helps them to come up with their new epicycle idea of 'dark energy' - a topic to be dealt with later. The Saga goes on.

<u>Stephen Hawking (1942 – 2018)</u>

No prominent scientist has ever been as shamelessly and ghoulishly exploited by the media as Stephen Hawking had been most of his life. Without knowing anything about him as a scientist, the public is in awe of his appearance, slumped forward in a wheelchair, talking haltingly through a metallic-sounding computer, conjures up images of an unearthly presence that can only communicate with us through a medium, like in a ghostly séance.

Jean-François Haït, a French science journalist has described him as: "*The modern Albert Einstein? The blockbuster author of unreadable books? The mind of a genius in an almost lifeless body… Beyond the myth, who really is Stephen Hawking?*"

Hawking's students described him to Hait as: "Hawking is like Delphi's oracle. We don t understand everything he says but we know it is meaningful….with him, we are lucky enough to tackle topics usually reserved for senior researchers. It is difficult but very gratifying. But all in all, working with him is not easy. He has such a strong personality".

Hawking is not completely innocent in this image created for him, because he often makes prophecies (many of them convincingly true) such as: "- If humanity survives the rise of artificial intelligence, the ravages of climate change and the threat of nuclear terrorism in the next century, it doesn't mean we're home free…we must … continue to go into space for the future of humanity…I don't think we will survive another 1,000 years without escaping beyond our fragile planet"

But, if we forget about the popular image and pitiful appearance, we can start evaluating him as a scientist. He was born in Oxford, England, during the war and his parents chose that city because the German Luftwaffe and the British Royal Airforce had a gentleman's agreement that the Germans won't bomb Oxford as long as the British left Heidelberg and Gottingen alone (the working-class peons had to take their chances in industrial districts, as long as the elites were safe and snug in their intellectual cloisters).

In 1959 Hawking went to the University of Oxford on a scholarship and started his Ph.D. studies under Professor Dennis Scima (Fred Hoyle, as his first choice, was too busy already to take him on). Shortly after he started his cosmological research under Scima, he was diagnosed with ALS – an incurable motor neuron disease, the source of the affliction he became famous for. Doctors gave him a maximum of two years to live, which understandably threw him into a deep depression.

In 1963 he met his future wife, Jane Wilde, a university student at Westfield College in London, studying languages. She knew about his condition and the dire prognosis, but she had deep 'motherly' feelings toward him and saw him as someone she could care for. As she said: "I wanted to find some purpose to my existence". They were married in 1965 after Hawking finished his Ph.D. and got a research fellowship in theoretical physics at Gonville and Caius College in Cambridge. His first research project focused on black holes.

By the time Hawking started his research much was already known about stars, including many old, dead stars. Some dead stars that have been compressed to the size of a planet and had a density of millions of grams of mass per cubic centimeter were called *white dwarfs*. Further analysis of Einstein's general relativity equations suggested that if a star such as our Sun were compressed to the radius under 9 kilometers, its gravity (which depends only on mass and diameter) would trap everything, including light, forever. John Wheeler, an American physicist dubbed them 'black holes' in 1969.

Further calculations by busy theoreticians suggested that it was possible to reach densities far greater than that of a white dwarf if the star was squeezed even further by forcing the electrons to combine with protons in the atomic nucleus. It would collapse the atom, which is mostly empty space between the nucleus and the electrons. The star would turn into a '*neutron star*' with a mass 20% higher than our Sun. Calculations also showed that any neutron star with a mass of 3 solar masses would collapse, under its own weight, into an object with the size of zero diameter, called a 'singularity'.

The physics of gravity for any material body collapsed to a small enough space is quite straightforward, even with Newtonian mechanics. As is described in *Humane Physics,* escape velocity from a star is defined as the speed required to leave a star fast enough, so that its speed would be reduced to zero only at 'infinity, so it would never fall back. The object's total energy (kinetic plus potential) would be zero according to:

$$TE = mv^2/2 - GmM/R = 0$$

where 'v' is the initial speed; 'm' is the mass of the object; 'G' is the gravitational constant; 'M' is the mass of the star and 'R' is the diameter of the star. So, rearranging the equation we get the equation for the escape velocity:

$$V_{esc} = (2GM/R)^{1/2}$$

From this formula we can see that the smaller the star's radius 'R' , the greater the escape velocity. The maximum escape velocity (according to the Relativity theory) is the speed of light 'c', so the equation for the black hole radius is:

$$R = 2GM/c^2$$

This radius is called the '*Schwarzschild-radius*' named after a German astrophysicist. For our Sun this radius is 3km. That's quite a reduction from the original radius of roughly 700,000 km. Since this equation can be applied to *any* object, physicists calculated this 'Schwarzschild-radius' for Jupiter (~3 meters); Earth (~9mm); moon (0.1mm), etc. The surface at the Schwarzschild radius is called an *event horizon*. General relativity tells us everything about black holes except the situation at the center where the density becomes infinite (called a *singularity*). Physicists hate infinites because they are not real numbers that calculations could be based on.

Hawking started collaborating with Roger Penrose, a mathematician, 11 years his senior, whom Hawking met during his Ph.D. course, and who introduced the young physicist to the idea of spacetime singularity. During the original discussion, Hawking started musing aloud: "I wonder what would happen if you applied Roger's singularity to the entire Universe". That musing became the central idea of his Ph.D. thesis and provided further inspiration for the rest of his research career. They worked very well together, each bringing his strength to their work: Penrose was a highly skilled mathematician (Hawking's weak spot) while Hawking was a highly intuitive theoretician who could visualize what was happening inside the physical systems they were studying.

In the meantime, astronomers were busy all over the world and experimental and observational data kept accumulating. In 1963 Allan Sandage, an American astronomer, discovered *quasars*, stellar objects that produce a lot of radio waves. Most of these 'objects' were actual galaxies, but in 1963 American astronomer Maarten Schmidt identified an actual star: 3C 273 that emitted a lot of radio waves. To their surprise, they found that these waves were redshifted over 16%, beyond the usual redshift observed before. Other radio stars were discovered with similar large redshifts and, since astronomers didn't know what they were, they called them 'quasistellar objects' or 'quasars.

Then, in 1967 Susan Jocelyn Bell, an astrophysicist from Northern Ireland discovered '*pulsars*' a different kind of radio star that emits radio signals that pulse up to 3000 times per second with incredible precision. The first two pulsars discovered were called 'LGM 1' and 'LGM 2' jokingly referred to as "Little Green Men" because of their regularity, as if they were interstellar beacons set up by aliens. These turned out to be rotating neutron stars.

The next observational breakthrough came in 1965 when two young researchers at the Bell Laboratories, Arno Penzias and Robert Wilson, were testing an antenna they used as a radio telescope. They were troubled by a faint hissing noise, whichever direction they pointed their antenna. The amusing story often told, is that the two of them spent days crawling over their radio telescope to scrub off the pigeon droppings, trying to eliminate the annoying noise. This noise was due to electrodynamic microwave radiation that corresponded to a calculated temperature of 3K degrees. This was the theoretically predicted background radiation left over after the Big Bang, as George Gamow (Russian physicist) had calculated in the 1950s. This was the first real observation supporting the idea that the Big Bang model could actually be an accurate description of how the Universe started.

So, by the mid-1960-s, both the theory of 'black holes' and the 'Big Bang' theory seemed quite plausible. Conclusion: **if the general theory of relativity is a correct description of the Universe,** there must have been a singularity at the beginning of time.

It's ironic how every major cosmological theory is based on one experimental piece of evidence. The expanding Universe theory is based on Hubble's observation of redshifted galaxies and the Big Bang theory is supported by the background radiation – all further calculations are using Einstein's general theory of relativity. However, no experimental or observational data was yet directly connected to either of these theories. By 1990 physicists were 95% certain that an x-ray emitting object Cygnus X-1 actually contained a black hole.

During all these development Hawking was very busy with his life. In 1967 his first child, Robert was born, followed by Lucy in 1970. Then, in 1971, he published a book titled *The Large Scale Structure of Spacetime* with George Ellis, an old classmate from South Africa, a book so incredibly complex as to be completely unreadable for most who tried. Radio astronomer John Shakeshaft, when asked by Hawking if he liked it, replied" I thought I might make it to page 10 but I only got to page 4". As his reputation grew in scientific circles, Hawking also traveled increasingly, often to America, Europe and even to Moscow (in 1973)

Finally, his major breakthrough happened in 1974 when he published in Nature a paper titled: "Black Hole Explosions". It described a new idea, unbelievable at the time, which became known as "Hawking Radiation" or "Black Holes are not really black". By mathematical reasoning, Hawking came to the conclusion that tiny black holes could emit radiation and even explode.

The amusing story behind this breakthrough was about Hawking trying to disprove the theory of a young researcher, Jacob Bekenstein at Princeton University, who applied Thermodynamics to black holes – only to find out that Bekenstein was right.

According to quantum theory, predicted by Paul Dirac in 1928, electrons have antiparticles, called positrons, and the pair could be spontaneously generated from empty space, converting the energy, always present in space by Einstein's $E=mc^2$ formula. Then, a fraction of a millisecond later they annihilate each other with a burst of energy back into space. According to quantum theory, every particle has a corresponding anti-particle (which has been observed in experiments) so the 'empty' vacuum is in fact a seething cauldron of constantly generating and annihilating pairs of virtual particles.

According to Hawking's calculations, if the electron-positron pair is generated in the vicinity of a black hole, one of the pair might be captured by the black hole, and the other escapes into space as radiation. The black hole is slowly losing mass because the energy for the radiation has to come from the black hole. Large black holes will not shrink because they also capture mass from traces of gas and dust always present in the universe, but for miniature black holes the Hawking radiation could lead to the black hole 'evaporating'. When Hawking applied Thermodynamics in his own calculation, the temperature of this radiation would match the same predicted by Bekenstein.

The Hawking Radiation paper is the pinnacle of his career. He won accolades from his peers for having one of the most brilliant minds in science, but he never got a Nobel Prize because no one has yet proven his ideas. Although it became widely accepted, Hawking Radiation was never proved by astronomers or physicists - if it had been, it would almost certainly have earned him the Nobel Prize. In January 2016, Prof Hawking gave a Reith Lecture broadcast on the BBC in which he joked that his lack of a Nobel Prize was "a pity".

Maybe compensation for his lack of Nobel Prize, his fame in the general population reached 'astronomical' proportions after he published his famous book: "*A Brief History of Time: From the Big Bang to Black Holes*" in 1988, written for the general audience. To date, it has sold over 10 million copies and was translated to dozens of languages. Overnight Stephen Hawking became a 'cult figure' to the masses that always want something dramatic for their entertainment. In this case, a crippled genius who could be gawked at on TV as he is wheeled around in his chair, unable to speak but making an Oracle-like prediction about the fate of our world.

The reason I included him in the "Epicycle Physics" part of the book is the fact that almost all of his accomplishments are based on mathematical deductions with very little experimental evidence.

supporting them. The last few decades of his life he was musing/calculating about baby universes, speculation on the origin of the universe(s) all highly mathematical and absolutely unprovable by any means available today.

Some fans see deep religious significance of his ideas but Hawking himself became more and more outspoken against religion and any other form of mysticism which he called popular nonsense. As John Gribbin, one of his biographers wrote: "Through his work, Hawking's early agnosticism had become more overtly atheistic, and with his no-boundary theory he had effectively dispersed with the notion of God." This became a major conflict with his deeply religious wife and led to their separation in 1990.

He died in 2018, 59 years after doctors gave him no more than 2 years to live. He had stubbornly refused to succumb to his illness and the world was enriched by his genius.

Dark Matter, Dark Energy

Dark Matter

Fritz Zwicky, a Swiss astronomer working out of Caltech in California, was a very abrasive person who called most of his colleagues 'spherical bastards' because, as he said, "they were bastards any way you looked at them" (he was a proponent of super-symmetry theory). So much for academic collegiality. Nevertheless, he has made significant contributions to science. He suggested in the early 1930s that there might be additional, *invisible* matter permeating the universe that he called 'dark matter'. What led him to this startling suggestion was the results of his observation of galaxies in the *Coma galaxy cluster*.

It had been known for some time that all spiral galaxies rotate, completing a full rotation once every billion years. This was determined by observing the blue shift of spectra on one (approaching) side of the galaxy and a redshift on the opposite (receding) side. From the speed of rotation, astronomers could deduce the mass of the galaxy because the greater the mass, the bigger the total gravitational pull and, according to Newtonian physics, the faster the rotation of the individual stars around the galactic center.

However, what Zwicky found was that the stars on the edge of the galaxy were moving too fast for the estimated mass of the galaxy to keep them in its gravitational pull and those stars would be flung free like "water droplets thrown off a spinning bicycle tire" as Brian Green explained in his *The Fabric of the Cosmos* book. According to Zwicky's calculation, only a gravitational pull *400 times greater* than the estimated mass of the galaxy could prevent that.

Further observations by astronomers during the next 30 years, by Vera Rubin of the Carnegie Institution, in particular, arrived at a similar conclusion – the galaxies must be immersed in a giant ball of invisible matter that would provide the needed gravitational pull. The galaxies and stars are like "beacons floating in a giant ocean of dark matter". Rubin's calculations were supported by radio astronomers who could measure the speed of gas clouds beyond the visible spiral arms. The farther these clouds were from the galactic center, the faster they would rotate until they reached a maximum value, after which their speed should decline according to Kepler's law of planetary motion. What they found, to their surprise, was that the speed did not decline but resulted in what now is called a "flat rotational curve". This flat curve has been observed in almost 1000 spiral galaxies. Today it is considered the best evidence for the existence of dark matter.

So, what is this hypothetical dark matter made of?
Nobody knows.

Speculation about WIMPs (Weakly Interacting Massive Particles) or MACHOs (Massive Compact Halo Objects) have been proposed but so far no experiments resulted in actual observation. Wimps could be a little bit like 'heavy' neutrinos and MACHOs could be large gas planets or unlit stars. Or it could be something else, or 'dark matter' is just a fancy name for "we don't know".

That's what science and cosmology have been reduced to – unsupported speculation.

Dark Energy

The last 'dark' piece, estimated to be 70% of the cosmology pie, is what has been called *'dark energy'*. In 1916 Einstein was dismayed to find that the equations he had developed in his General Theory of Relativity contradicted the prevailing assumption of the universe being static and stable. The universe could collapse under its own gravitational force which would eventually pull all the stars and galaxies back together in a 'Big Crunch' - maybe into the state, it was in prior to the Big Bang. So, Einstein decided to play 'God' and 'disallow' this consequence predicted by his own theory. He did this by introducing an arbitrarily selected constant into his equation: *the cosmological constant* that would restore the universe to a comfortable static and stable state. No expansion, no contraction, just the way we like it. When Hubble's measurement contradicted this static theory, Einstein reluctantly removed the constant from his equation, calling it the "biggest blunder" of his life.

Once the constant was removed, physicists realized that there was room for the dynamically evolving universe, either contracting or expanding in size. They began in earnest trying to measure the actual speed of the universe's movement, primarily looking for deceleration of the expansion, which was the most logical thing to look for - after all, gravity was the only force they were aware of that could affect stars and galaxies. In order to accomplish this, they needed to measure the distance of the galaxies from us and the speed at which they are receding. And they had to do it for different time periods, which was made possible by looking at galaxies and stars at different distances from us because different distances implied different times in the past due to the time light needs to reach us.

The speed of recession is the easier value to find by the well-established method of measuring the doppler shift in the galaxy's spectra. That's exactly the method Hubble used to determine his law, according to which the farther a galaxy is, the faster it recedes from us. Hubble used the Cepheid variable stars to calculate distances of 'nearby' objects, such as the Andromeda galaxy, but this method is not reliable for very distant objects that astronomers needed to deal with if they wanted a clear answer for the whole universe. They needed a new '*standard candle'.* That's where supernovas entered the picture.

It was known that when a star exhausts its nuclear fuel, its gravitational force causes the star to collapse, causing its temperature to rise catastrophically, which leads to a gigantic explosion that, for a few weeks, can outshine the whole galaxy. One type of supernova, called *type Ia* was found to always shine with the same intensity, regardless of its size and distance from us. These supernovas all have the same critical mass, 1.4 times the size of our sun, hence the same intrinsic brightness. That's what makes them useful to be used as *standard candles.*

In the 1990s two research groups, at Berkeley National Laboratory and the Australian National University, managed to find four dozen *type Ia* supernovas in distant galaxies. At last, they could make their measurements and calculations. Repeating these measurements for different supernovas, at different distances from us, therefore in different times in the past, they arrived at a startling result: the universe is not only expanding, but this *expansion is accelerating* over time.

We have come full circle back to Einstein's cosmological constant. Clever physicists realized that if they put this constant back into the equation, *with a carefully selected value*, then the equations would *predict* the exact rate of acceleration that the astronomers measured. Since the constant itself had the role of 'hidden energy' in the equations, they dubbed this hypothetical new entity 'dark energy'.

How about that for epicycle physics?

Simon Singh, in his book *The Big Bang,* comments wryly that "every flawed model can be saved by such fiddling around with numbers".

Dark Physics

I have a confession to make: Even though I have repeatedly expressed doubts about the more esoteric theories of modern physics, I have to admit that any or all of them could prove to be descriptions of reality. My problem is not with the conclusions that physicists have reached out with the method they used to arrive at them.

And it is not ever about the speculative nature of their reasoning. The history of physics is full of speculations that later turned out to be proven correct. My problem is threefold:

-   Most of the time there is no experimental data to suggest that the speculation was a reasonable one.
-   By the very nature of the speculation, according to current estimates of present and foreseeable future technology, the theory will remain both unprovable and unfalsifiable.
-   Even if both 1. And 2. could be overcome in time, the speculations are totally irrelevant to humanity's self-interest because no conceivable application can be imagined for any of these theories. It feels more and more like speculation about the number of angels that are able to dance on the head of a pin.

When we review the heroic age of Physics – what now is called "Classical Physics'- we can't help seeing how our lives were enriched both materially and spiritually by discovering basic laws of our physical world. Once we got into modern physics, the benefits of this science started receding farther and farther. While Atomic Physics and Quantum Mechanics produced beneficial discoveries, especially in medicine, it also had a sinister side effect in the development of nuclear weapons. Once we arrived at cosmology, the theories and the experimental data could only tickle the curiosity in people who wanted to understand what the universe is made of and what rules it obeys. Nothing wrong with curiosity, of course, but I am troubled by the unbelievable contortions that physicists performed to come up with answers where no natural answer presented itself.

As theoretical physicist Alexander Unzicker wrote in his book: *Bankrupting Physics – How Today's Scientist are Gambling Away Their Credibility*: "Today, the major part of theoretical physics has instead gotten lost in bizarre constructs that are completely disconnected from reality, in a mockery of the methods that grounded the success of physics for 400 years".

Or, as a founding member of the Perimeter Institute Lee Smolin wrote in his book *The Trouble with Physics – The Rise of String Theory and the Fall of Science* "The current crisis in particle physics springs from the fact that the theories have gone beyond the standard model in the last thirty years fall into two categories. Some were falsifiable and they were falsified. The rest are untested - either because they make no clean predictions or because the predictions they make are not testable with current technology".

It's worth repeating what I wrote in *Book I – Classical physics* of the first volume: *Humane Physics*, chapter *The Limits of Science*:

"Here are a few quotes from leading physicists, mathematicians, philosophers, and other scientists (many of them Nobel laureates) of the last few decades.

Sheldon Glashow (Nobel Laureate, Physics, 1979) "Science is certainly slowing down....for the first time since the Dark Ages we can see how our noble search may end, with faith replacing science once again."

Steven Weinberg (Nobel Laureate, Physics, 1979) "...nor any other earthly accelerator could provide direct confirmation of a final theory; physicists would eventually have to rely on mathematical elegance and consistency as guides....might not reveal the Universe to be meaningful in human terms...all our 'whys' would eventually culminate in a 'because'."

Hans Bethe (Nobel Laureate, Physics, 1967) when asked "Could there ever be another revolution in physics like the one that accompanied quantum mechanics"? – "That's very unlikely".

Richard Feynman (Nobel Laureate, Physics, 1965) "The age in which we live is the age in which we are discovering the fundamental laws of nature, and that day will never come again. **..There**

will be the interest of the connection of one level of phenomena to another – phenomena in biology and so on" (my emphasis)

Robert Oppenheimer ("the Father of the Atomic Bomb", Princeton,1965) "The general notions about human understanding... which are illustrated by discoveries in atomic physics are not in the nature of things wholly unfamiliar, wholly unheard of, or new. Even in our own culture, they have a history, and in Buddhist and Hindu thought a more considerable and central place. What we shall find is an exemplification, an encouragement, and refinement of old wisdom".

Werner Heisenberg (Nobel Laureate, Physics, 1932) "The great scientific contribution in theoretical physcs that has come from Japan since the last war may be an indication of a certain relationship between philosophical ideas in the tradition of the Far East and the philosophical substance of quantum theory".

Niels Bohr (Nobel Laureate, Physics, 1922) "The search for the ultimate theory of physics might never reach a satisfying conclusion; as physicists sought to penetrate further into nature, they would face questions of increasing complexity and difficulty that would eventually overwhelm them.... For a parallel to the lesson of atomic theory ... [we must turn] to those kinds of epistemological problems with which already thinkers like the Buddha and Lao Tzu have been confronted, when trying to harmonize our position as spectators and actors in the great drama of existence"

Dreams and nightmares

Carl Sagan (1934 – 1996)

Carl Sagan could have become a Nobel-prize scientist, making huge steps forward in physics, astronomy, even biology, if he had followed the established route to academic success. He didn't choose this route because he was interested in 'big picture' subjects: space travel, life on other planets, peaceful co-operation on Earth.

He imagined a cosmos teeming with life, intelligence, and countless advanced species who have already solved their primitive evolutionary problems of domination and competition. In the movie "Contact", based on his science fiction novel of the same title, when Ellie Arroway is about to leave in an inter-galactic transport to meet a more advanced species, she is asked: "If you could ask them only one question, what would it be?" – Her answer was: "Well, I suppose it would be, how did you do it? How did you evolve, how did you survive this technological adolescence without destroying yourself?"

Carl Sagan was convinced that such a meeting could take place one day. He did everything his brilliance and energy enabled him to do to be ready for it. He contributed substantially to planetary science, to 'origin of life' research, to the planning of space missions. He wrote books on the meaning of science, on human evolution, and the dangers of irrational thinking and pseudo-scientific trends. He popularized science through mass media and promoted an image of human greatness in discovery, rather than conquest. He fought the establishment in their mad rush toward nuclear annihilation by pointing out the dangers of 'nuclear winter' that might follow even a limited atomic war.

He died at the relatively young age of sixty-two, in 1996 when a real possibility of nuclear disarmament and international co-operation seemed at hand. He thought that after half a century of cold war madness it was time to hope again. His main theme all through life had been: "what if?" and "why not?" What if it were possible to live in peace? What if it were possible to establish human roots on other planets, orbiting other stars? What if one day we could wake up from our nightmare of primitive savagery and learn 'just' to live and experience all the marvels of the universe? This "what if?" attitude may seem naïve to many 'serious' people who forget that telephone, radio, TV, airplanes, computers were all "what if?"-s that seemed impractical at one time or another.

In his book The *Pale Blue Dot*, Sagan describes the view of Earth from Voyager1 at a distance of 6.4 billion kilometers, as it was leaving the solar system in 1992. This was the first time in human history that we actually could see our home from outer space.

Though Carl Sagan judged it scientifically inaccurate, the TV series *Star Trek* by the visionary Gene Roddenberry represents the possibility of human greatness and cooperation for millions around the globe. That vision of the future is expressed in Captain Picard's speech to a group of earthmen from the 20th century, who had been in suspended animation. He told them that money had no meaning in the 24th century, and the only challenge was realizing one's creative potential (Star Trek The Next Generation, Season 1 Episode 26 "The Neutral Zone"). This assumption of peace, cooperation, the creative drive of unified humanity is central in Star Trek and expresses the dreams of billions on this planet. The reward would be the marvels of the Galaxy suddenly open and available to us.

Why is it taken for granted that only nightmares are possible in the human saga? To quote from *Lucifer's Hammer* by Larry Niven: "Dreams sometimes come true, even if you tell them".

The Physics of Star Trek

An excellent book on the TV series is *The Physics of Star Trek* written by Lawrence M. Krauss, professor of physics and astronomy. Professor Krauss does not dismiss the possibility of practical space travel, or the technological marvels Star Trek fans take for granted. However, he warns about claims that are scientifically impossible and he estimates probabilities for those that are not outright forbidden by natural laws.

The crucial question regarding space travel outside the solar system is the time it takes to get to even the nearest star. Alpha Centaury is 4 light-years away. Considering that light can circle our planet's equator about seven times in one second, the distance that light can travel in four years is humbling: 883,008,000 (eight hundred and eighty-three million, eight thousand) times the length of Earth's equator (4 x 365 x 24 x 60 x 60 x 7). The space shuttle can circle the planet once in 86 minutes (1.43 hours). At 28,000 km per hour, it would take about 144143 years to complete the journey to Alpha Centaury. Not a very practical proposition.

What if we could accelerate to near the speed of light? Couldn't we get there in a little over four years? Not if we wanted to stop by for a visit, because we would have to spend time to accelerate up to our cruising speed and an equal amount of time decelerating to zero speed. Assuming standard acceleration of 1 Earth-g, it would take about one year to accelerate to 77%, 2 years to accelerate to 97%, and 5 years to accelerate to 99.99% of the speed of light. Could it be done?

Using the still non-existent nuclear fusion motors of the impulse engine, according to Krauss's calculations:

> "Each time the *Enterprise* accelerated to half the speed of light, it must burn 81 TIMES ITS ENTIRE MASS in hydrogen fuel…Rocket-propelled space travel through the galaxy at near light speed *is not physically practical,* now or ever!"

Einstein's 'No' and Maybe'

It was Albert Einstein who first stated the Special Theory of Relativity that doomed any hope for faster-than-light travel, but it was also Einstein who, in his 'General Theory of Relativity', offered that "glimmer of hope".

Einstein's 'No'

We grew up with what is called in physics the 'Galilean transformation'. In essence, it says that speeds add up. If a train moves at 60 km/h relative to the ground, and I walk on the train in the same direction at 5 km/h, then I fully expect my speed relative to the ground to be 65 km/h. This is basic common sense.

So it is with sound. Both experiments and theory in physics agree that if I travel at 100 km/h toward a stationary fire-truck, blaring its siren at the usual speed of sound at 700 km/h relative to the air, the sound hits me at almost 800 km/h. No surprise here. On the other hand, if I am stationary and the fire truck is moving, the sound hits me at 700 km/h – the movement of the sound source does not make any difference, because the sound travels through air at its usual speed, and if I am stationary relative to the air, the sound arrives at my position with the same speed it has been traveling.

The surprise comes when we look at light. The speed of light is always the same for everyone, regardless of how the source of the light and the observer are moving relative to each other. Light is special, it does not care about common sense. It has its own rules. The essence of Einstein's Special Theory of Relativity can be understood from an experiment performed by American physicist Albert Michelson in 1881, suggesting that the speed of light is the same for everyone, regardless of how fast the source of light, or the observer, moves. *Exactly* the same!

Now, this is really weird. If a beam of light is projected from the moon at its usual speed in a vacuum of 300,000 km/s at rocket moving toward the Moon at 100,000 km/s, the observer on the moon could expect the light to arrive at the rocket at 400,000 km/s. Yet it doesn't. The occupant of the rocket measures the speed of the arriving light beam at its usual speed of 300,000 km/s - *the exact same speed* as if the source of light were standing still relative to the rocket!

The rocket's pilot, measuring the speed of the rocket relative to the moon concludes that light must have left the moon at 200,000 km/sec, yet the observer on the moon measured the light beam leaving at full speed. Which of them is right? Both and neither, because their measurements of distance and duration are not the same.

Experiments with the speed of light were performed countless times, with ever-increasing accuracy, because physicists hoped that the experimental results were due to some kind of error. No such luck, the experiments proved this shocking behavior of light correct every time.

When Einstein worked out his theory, based on the invariant nature of the speed of light, he came up with transformation formulas that tell us exactly how the two observers measure distance (space) and duration (time), depending on their relative speed.

According to this formula: if platform 'B' (e.g. a space ship) is moving relative to another platform 'A' (e.g. Earth) at speed v_1; and a bullet is fired from platform 'B', in the same direction, at speed v_2, (relative to 'B') than the bullet's speed is measured from platform 'A' as:

$$v = (v_1 + v_2)/(1 + v_1 v_2 / c^2)$$

where 'c' is the speed of light in vacuum (300,000 km/s).

If both speeds were equal to the speed of light, then adding the two velocities would yield:

$$(c + c)/(1 + (c\, c)/ c^2) = 2c/ (1 + 1) = c.$$

Exactly the same, as if at that point I could not add even a tiny fraction to the speed of the bullet (beyond the speed of my spaceship) Surprise! Nothing seems to be able to go faster than the speed of light!

This 'seems to' becomes a 'for sure' when we look at one of the consequences of having to redefine space and time: the fact that the mass of material objects increases with speed. The formula for the mass increase uses a correction factor similar to what we used in our speed addition formula:

$$m = m_0 / (1 - v^2/c^2)^{1/2}$$

where m_0 is the mass of the object measured by the person at rest relative to it, 'v' is the relative speed of the two observers and 'c' is the speed of light. At very low speeds (when v^2/ c^2 is tiny) the $(1 - v^2/ c^2)$ correction factor becomes so close to one that, for all practical purposes, we all measure the same mass for the same object, even if we move (uniformly) relative to each other.

At higher speeds, however, the mass of the same object will be measured considerably more by the observer who is moving relative to the mass, compared to the value measured by the one who is at rest relative to it. And this fact is the reason why we can not have faster than speed-of-light travel.

Viewed from Earth, as the rocket gains speed, its mass increases and therefore it will require greater and greater force to increase its speed by the same amount. Eventually, as the rocket's speed approaches the speed of light, the correction factor by which we have to divide the mass will approach zero, thus the mass approaches infinity. Therefore the force the rocket needs to increase the speed even to reach the speed of light will also approach infinity. Since the rocket doesn't have access to infinite forces, the rocket will never quite reach the speed of light, let alone exceed it. This fully proven theory is Einstein's 'No' that so discouraged science-fiction fans and writers all over the world.

Einstein's 'Maybe'.

Just as the "Special Theory of Relativity" said 'No' to galactic space travel, Einstein's "General Theory of Relativity" left the door open for a 'Maybe'.

The first theory was called "Special" because its validity is restricted to "Inertial" systems where Newton's law of inertia is valid. Such systems are often said to be moving at uniform speed, but since we can never find Newton's 'absolute space', this uniform speed remains undefined.

Einstein developed a theory that would be valid in any system, inertial or otherwise. While the mathematical proof of the "Special Theory of Relativity" never exceeds high-school algebra, proving the "General Theory of Relativity" requires very complicated reasoning and advanced tensor-calculus that took Einstein over ten years to develop.

I will outline the concepts and some of the conclusions to indicate why Carl Sagan's (and *Star Trek* fans') dream is not necessarily impossible. The basic idea came from what Einstein later described as "the happiest thought of my life". He realized that, without looking out of a closed box, we could not tell whether we were freely floating in intergalactic space, totally free from any gravitational effect, or freely falling in a gravitational field of some huge mass, like a planet or a star. He called this insight the "Principle of Equivalence".

This equivalence is exactly what enables NASA engineers to simulate zero-gravity environment by a freely falling spacecraft (a modified Boeing 727, called the "vomit comet") in which, for a short period of time (about 30 seconds), the training astronauts can't tell without looking outside that they are falling toward Earth and not in zero-gravity space.

Einstein used the local validity of the Special Theory of relativity to develop his field equations that give a new interpretation to Newton's assumption of "gravitational force".

Newton could not explain how this "action at a distance" of gravity operated, how it could reach across millions of miles of vacuum-filled space and exert such an incredible force on massive objects like planets to make them constantly change direction and follow an elliptical path.

According to Einstein's general theory of relativity, there is no force at all. The sun doesn't know about the planets, the planets don't know about the sun. They don't pull on each other; there is no attraction at all. What each chunk of matter does, it does it by itself, independently of any other chunk, but the effects add up and create the reality of our planetary system.

We have already seen from the special theory of relativity that space and time are not absolute, because observers measure distances and durations differently from one another, depending on their relative uniform motion. As it turns out, it is not only relative speed that influences our space and time: matter and energy (interchangeable, as we will see later), as distributed through space, will also have an effect.

This effect will redefine the concept of "straight line" and "shortest distance" between two points in such a way that in the vicinity of large masses like the sun, the straight line will become distorted into a curve. So the planet 'thinks' it is going in a straight line, minding its own business, unaware that the sun is there, yet it goes around in a closed loop. Very much like a man going north in a straight line on Earth will have to follow a curved line through the two poles, leading back to the point where he started.

Every object follows its motion through space along the shortest, straightest path, as it is distorted in the vicinity of big lumps of matter. The sun does not affect the planets; it affects the space around itself. John Archibald Wheeler said, "Matter tells space how to curve; space tells matter how to move".

For *Star Trek* fans, this distortion effect provides the 'maybe' for faster-than-light travel. We hope to find shortcuts (wormholes, black holes) through the folds of distorted space, that could allow us to "jump" huge distances, not by following the surface of the fold but by cutting through it at one point and re-emerging at another, light-years away. Science, as we understand it today, does not preclude this possibility.

The Ethics of Science

So many things happened in my life that I could hardly believe, even as I was experiencing them.

Two episodes stand out. The first was Neil Armstrong's "small step" on the moon, forty years ago. It represented everything we had ever hoped for, dreamed of, and didn't quite dare to expect. A step leading us to the stars and maybe human survival.

The second was the 9/11 attack on the World Trade Centre. I felt as if I were witnessing someone walking over my grave. In those images of exploding planes and collapsing buildings, I saw the potential destruction of the human species.

Both events happened, both were a huge step in alternate directions: triumph and annihilation.

One day we may look up at the sky and see the first alien starship visiting us from another civilization. One day we may pick up an intelligent message from outer space, just as Carl Sagan imagined in *Contact*. One day we may even invent our own intergalactic space vehicle and go for a look.

On the other hand, one day we may look up and see a mushroom-shaped cloud, for a fraction of a second, before we lose our eyes and our lives. There is a very good chance that on that day mankind will finish the job started at Hiroshima on August 6, 1945, and put an end to the human saga. On that day, if there is a God, he will weep. It will have been such a horrible waste!

Which of the two alternate futures awaits us depends, to a large degree, on the scientists themselves. Not just on the breakthroughs they provide, leading us to a possible glorious future but, more importantly, on the ethical stand they take when bribed or coerced by madmen to produce weapons of mass destruction.

Without the greatest brains of the twentieth century (Einstein, Bohr, Oppenheimer, Feynman, Teller, etc.) political leaders would be more or less harmless fools. They might mess up everything they touch but could not threaten all of humanity and every living species on the planet with extinction. It was scientists who gave them the tools of total annihilation.

No "ethics of science" course was ever taught to me as a physics student at university. Medical students now grapple with the ethics of medicine as it manifests in subjects such as cloning, euthanasia, abortion, stem-cell research, and genetic manipulation. They are given a philosophical & legal framework in which to consider these issues, and guidelines for coping with them.

Nothing highlights the issue of ethics in science as well as the dilemma confronting scientists in the Manhattan Project. The justification for developing a nuclear weapon seemed overwhelming. None of those great scientists contributing to it were evil or greedy monsters. What they were is unbelievably naïve. We can see from their writings after the atomic bomb became a reality and they felt horrified by what they had done and scrambled to influence national policy in a positive direction.

After the Nagasaki bombing, members of the Scientific Panel – Lawrence, Oppenheimer, Compton, Fermi -- wrote a letter to Henry Stimson, Secretary of War on August 17, 1945:

"The development, in the years to come, of more effective atomic weapons, would appear to be most natural element in any national policy of maintaining our military forces at great strength; nevertheless we have grave doubts that this further development can contribute essentially or permanently to the prevention of war."

Robert Oppenheimer expressed his fears more forcefully on his last day as director of the Manhattan Project:

"If atomic bombs are to be added as new weapons to the arsenals of a warring world, or to the arsenals of nations preparing for war, then the time will come when mankind will curse the name of Los Alamos and Hiroshima."

Some of the scientists disagreed. Edward Teller, who later led the effort to develop the hydrogen bomb, wrote:

"The reason they gave just made me mad…The important thing in any science is to do the things that can be done. Scientists naturally have a right and a duty to have opinions. But their science gives them no special insight into public affairs."

This opinion was expressed, much more forcefully, by Nikita Khrushchev to Andrei Sakharov, father of the Russian hydrogen bomb, who later became a prominent dissident, when Sakharov recommended suspension of further tests:

"Sakharov writes that we don't need tests…He's moved beyond science into politics. Here he's poking his nose where it doesn't belong. You can be good scientist without understanding a thing about politics…Leave politics to us – we are the specialists. You make your bombs and test them, and we won't interfere with you; we'll help you." (Andrei Sakharov: "Memoirs")

Oppenheimer wrote about his feelings after the successful first test:

> "We waited until the blast had passed, walked out of the shelter and then it was extremely solemn. We knew the world would not be the same. A few people laughed, a few people cried. Most people were silent. I remembered the line from the Hindu scripture, Bhagavad-Gita: Vishnu is trying to persuade the Prince that he should do his duty and to impress him he takes on his multi-armed form and say, "Now I am become Death, the destroyer of worlds". I suppose we all thought that, one way or another."

There are many reasons why scientists decide to participate in weapons development. Here are a few:

- We need it to defend our nation.
- These weapons will prevent war.
- The enemy won't stop, so we can't either.
- If I don't do it, someone else will.
- It is a great challenge and "super physics".
- Nobody else would fund my research.
- It's the only job I could find.
- It is fun!

A few of these arguments sound convincing and did convince superb minds back in the 1940s to develop nuclear weapons for the first time in human history. What many scientists don't realize is that anything can be justified by clever demagogues, if they pick their facts and reasons carefully, omitting anything that contradicts their pre-determined conclusions.

What scientists should never forget: the arbiter of any theory is an experiment. Arguments aside, the world did not face total destruction before nuclear weapons were developed. Now it does.

As Richard Rodes writes in *The Making of the Atomic Bomb*:

> "...the death machine that we have installed in our midst will destroy the nation-state, ours and our rival's along with most of the rest of the human world. The weapons with which the superpowers have armed themselves – collectively the equivalent of more than one million Hiroshimas – are linked together through their warning systems into a hair-trigger, feedback-looped contrivance, and no human contrivance has ever worked perfectly nor ever will. Each side is hostage to the other side's errors. The clock ticks. Accidents happen."

Nothing will change this fact! Yes, the war *might* have lasted longer without it. Yes, there *might have been* another world war or two without the nuclear deterrent, instead of the many, many small wars all over the world that have killed millions since WWII. But humanity would not face the possibility of extinction today if scientists refused to participate in that 'superb and magnificent' project of evil and insanity.

The ethical precept that should be taught to science students all over the world should be the same as that of the Hippocratic Oath for medical students: "First, do no harm!" Say no to weapons research, say no to projects that harm the environment, that cause pain and suffering to life on this planet. Nothing can be simpler.

Resolving conflicting loyalties

The suggestions I am making in this chapter should be read as guidelines that I have found useful in my own life. No one can follow them with absolute perfection, because human beings have conflicting motivations: what Edward O. Wilson called individual-level selection and group-level selection in our evolutionary process (*The Meaning of Human Existence*). The result of individual-level evolutionary selection predisposes us to favor our own and our progeny's survival over the interests of our group. The result of the group-level evolutionary selection motivates us to serve the interests of the various groups we are part of. As he so eloquently states:

> "We are unlikely to yield completely to either force as the ideal solution to our social and political turmoil. To give in completely to the instinctual urgings born from individual selection would be to dissolve society. At the opposite extreme, to surrender to the urgings from group selection would turn us into angelic robots - the outsized equivalents of ants."

With these caveats, I will attempt to define human morality in a logical and systematic way that should serve as a compass for future scientists when they struggle with the conflicting loyalties that they will unavoidably encounter.

Only the first three reasons cited above for weapons research deserve further analysis. These were the reasons that compelled well-meaning scientists to develop nuclear weapons during World War II. All three expressed, in different ways, their loyalty to their country.

The human species is a tribal species, just like wolves and gorillas. We depend on one another for survival. The question of loyalty to our tribe often conflicts with our other loyalties: to family, humanity, religion, etc.

The relationship of our social concepts can be seen as follows:

1. We have evolved with nearly identical needs for survival.

2. Our nearly identical needs created nearly identical values.

3. Our nearly identical values created a set of ethical rules (dos and don'ts)

4. Our dependence on one another created a need for loyalty to our ethical rules.

5. Our loyalty to ethical rules created an unwritten social contract apart from the laws of the land as defined by the ruling elite. Those laws are specific to one culture or one nation-state. The unwritten social contract recognizing human interdependence is universal. All cultures through history have known that murder and theft are wrong. Awareness of the rules of the social contract is called our 'conscience', or knowing right from wrong. This universal concept of 'right conduct' is called morality.

6. The unwritten social contract created standards of socially acceptable behavior. Any act or attitude that enhances the chances of survival for the group is good. Any act or attitude that harms the chances of survival for the group is bad. Since individual members accept the protection and nourishment of the tribe, the only moral conduct is to seek individual survival/welfare only through the survival/welfare of the tribe. If the two are in conflict, the needs of the tribe come first. We call those who consistently demonstrate their willingness to defend the tribe, even at a great personal sacrifice, 'heroes'. Those who betray the tribe we call 'traitors' and treason is usually punishable by death or expulsion.

7. In our complicated world, individuals have simultaneous and often conflicting memberships in many groups: immediate family, extended family, friends, neighborhood, school, work,

religious denomination, political party, social organizations, nation, race, gender, species, and life.

8. Resolving conflicts requires prioritizing our loyalties.

9. Since a sub-group accepts the protection and nourishment of the larger group of which it is a part, the only moral conduct is to seek survival/welfare of the sub-group ONLY through the survival/welfare of the containing group. If the two are in conflict, the needs of the containing group come first.

10. In this sense, our ultimate loyalty should be to life. Life on this planet is the ultimate containing group. We are all part of it. It nourishes us all. If we betray it, if we destroy it, we destroy ourselves.

Morality is about the survival of the whole of which we are part. At Nuremberg, the claims of loyalty to the country did not excuse crimes against humanity. There should be 'crimes against life' trials for those destroying ecosystems.

I felt ashamed during the first Gulf war when the TV showed us the pelicans covered in oil. I felt that 'we' had betrayed our common heritage. I felt the need to apologize to the cormorants, to all animals, to Life.

Morality has always been in human consciousness, though not always verbalized, defined, analyzed, explained, but lived by a sufficient number of the tribe to assure survival. Tribes that failed the test of morality died and disappeared.

Morality is the prerequisite of survival. Nature created us. We are an inextricable part of it and have no choice but to behave by its rules. Morality is our interdependence expressed in thought and deed.

Morality is life-affirming. Immorality embraces death. Maybe not immediately, not personally, but the human species can die by many, many incremental steps. Destroying our habitat bit by bit will do it: the poisons in air, water, and food are all material manifestations of immorality, of some human being, somewhere, in some capacity, failing the test of ethical behavior.

We have to sort out our loyalties in a way that doesn't destroy us. Each containing group takes precedent. My loyalty to my country has to take second place behind my loyalty to humanity. And my loyalty to my species has to come behind my loyalty to universal, interconnected, miraculous, and fragile life we are all part of.

It could take one dumb asteroid to destroy us. It could take one dumb humanity that developed too much power before developing enough sense. Morality could save us from that fate.

Random House defines the word 'honor" as: "high respect as for worth", or "honesty or integrity in one's belief and actions". 'Honourable' is defined as "worthy of honor and high respect". An honorable man is someone who follows the universally accepted rules of right and wrong and, as a consequence, is admired by human beings everywhere. Gandhi was admired around the world, even though he was treated like a criminal by the British ruling class.

The word 'honor' has been hijacked and co-opted by the elite that holds most of the wealth and power, and its primary motivation is to maintain this position. Honour came to mean 'loyalty' to whatever group, standing for whatever goal or principle. German officers' sense of 'honor' prevented them from standing up to Hitler. However, we all understood why John Le Carre named one of his best novels "The Honourable Schoolboy", even though Jerry Westerby betrayed his masters.

'Honour' does not mean loyalty. SS guards had loyalty. It does not mean 'integrity'. Bin Laden had integrity. His belief in his horribly misguided crusade seemed genuine.

Honour is the highest praise among human beings. A judge is called 'your honor' because he is supposed to have the wisdom and integrity to represent our best interests. Honor means representing this interest. A secret agent, pretending and lying in order to defeat evil from inside is an honorable man. A law-abiding citizen in an evil regime is a dishonorable human being.

Our social concepts are linked into a cause-and-effect logical chain: survival – needs – values – ethics – social contract – morality - honor.

This chain ties honor to our survival needs, regardless of what our rulers pretend our interests are. Citizens know what their interests are, without being told. They want to be healthy, secure, productive; to raise their families in a wholesome, peaceful, cooperative society. Most don't believe they need to send their sons and daughters to the other side of the globe to kill and be killed.

And, most important to the readers of this book: scientists have a very special moral obligation to humanity. No matter what the justification, do not help immoral leaders acquire the tools they need to force their will on the citizenry. Do not participate in weapons development and do not work for industries that damage the environment. If you do, you will betray the highest loyalty: to life, including yours and your loved ones.

Your Math toolkit

This chapter summarizes what the reader needs to know about math and how to use it to follow the rest of the book. In the following pages, I will introduce the most common mathematical notations used in this book, together with the most frequently used formulas and equations. These rules are very easy to prove and in your old high school math book, you will find plenty of examples as to how these rules are used in specific cases. For my readers familiar with advanced high school math, this could be used as a short refresher summary.

Algebra

It is with the start of Algebra where most high school students fade out. The reason for that panic, I always thought, is going too fast, skipping over important steps required to see how Algebra is real and makes perfect sense. Just as Physics education should start with Astronomy, Mathematics education should start with numbers.

I am sure that you are wondering what do you need to know Algebra for. On one hand, you need it if you really want to understand Physics on a deeper level. On the other hand, quite often, knowing these simple algebraic formulas make your life easier when you try to solve simple problems in real life. I will give you 2 examples.

1./ Suppose you want to figure out the square footage of your house that is 44.8 feet by 45.4 feet (for the insurance company). You don't have your calculator on you and you want to multiply these two numbers. You can get a ballpark figure if you multiply 45 by 45. I will show you a simple trick to multiply any number that ends with the digit five, by itself (get the square of). The last two digits of the result will always be 25, preceded with the first number multiplied by a number one larger: in this case 5 – resulting in 2025. This trick will work for any number that ends with the digit 5. Check it out:

$$15 \times 15 = 225$$
$$25 \times 25 = 625$$
$$35 \times 35 = 1225$$
$$45 \times 45 = 2025$$
$$55 \times 55 = 3025$$
$$65 \times 65 = 4225$$
$$75 \times 75 = 5625$$
$$85 \times 85 = 7225$$
$$95 \times 95 = 9025$$
$$125 \times 125 = 15625 \; (12 \times 13 = 156)$$
etc.

It works every time. Why? I can prove it to you with simple algebraic deduction:

Any number that ends with the digit 5 can be represented as $10a + 5$, where 'a' stands for the number made up of the digits preceding the 5.
For example:

$$35 = (10 \times 3) + 5$$
$$75 = (10 \times 7) + 5$$
$$125 = (10 \times 12) + 5$$

So, if we perform the multiplication, step by step (multiplying every term in the first bracket with every term in the second bracket):

$$(10a + 5)(10a + 5) =$$
$$100a^2 + 50a + 50a + 25 =$$
$$100a^2 + 100a + 25 =$$
$$100a(a+1) + 25$$

Can you see it? The last two digits are 25 and the preceding digits (starting with the hundreds) are the first number multiplied by one larger. Great! So who needs a calculator if you don't happen to have one on you when you need it?

The second example I promised involves my neighbor who wanted to erect a flag pole in his front yard. He knew the height of the pole and he knew the distance of the anchor from the base of the pole, so he wanted to know the length of the rope he needed from the anchor to the top of the pole. He laboriously drew a schematic on graph paper, measured the length of the rope (the hypotenuse that is the longest side of a right triangle) on the graph, and converted this length to actual meters. Had he paid attention in high school when they were teaching Pythagoras's theorem, according to which the length of the hypotenuse of a right triangle is the square root of the sum of the squares of the other two sides, that is

$$c^2 = a^2 + b^2$$

He could have saved himself the trouble of graphing it all out, measuring, and converting back to actual meters. So, if you know that the length of the pole is 4 meters and the distance of the anchor from the base of the pole is 3 meters, then you will immediately know that the length of the rope you need is the square root of $4^2 + 3^2$, that is $\sqrt{25} = 5$: that is 5 meters. No problem at all.

These examples illustrate that knowledge of algebraic formulas is useful in everyday life, and absolutely necessary in order to understand the very Physics your life depends on every day. So how does algebra work?

Here is an example: think of two one-digit numbers, say 5 and 3. What happens when you multiply the sum of these 2 numbers with the difference of the same two numbers? Let's try:

$$(5+3)(5-3)$$

When you do the math, you will get 16 (from 8 times 2). You can get the same result by doing the following (multiplying every number in the first bracket with every number in the second bracket:

$$5\times5 - 5\times3 + 3\times5 - 3\times3$$

The two middle terms cancel out, so we end up with:

$$5\times5 - 3\times3$$

which is

$$5^2 - 3^2$$

so it seems that for the numbers of 5 and 3 the following is true:

$$(5+3)(5-3) = 5^2 - 3^2$$

that is: if we multiply the sum of 5 and 3 with the difference of 5 and 3, then we get the square of 5 minus the square of 3.

I suggest you try it for any two numbers and do the math: you will find that is true for any two numbers you can think of, small or large. And that is where Algebra comes in: it is a shorthand notation, using letters, instead of numbers:

$$(a+b)(a-b) = a^2 - b^2$$

This formula means that the relationship expressed by it is true for any two numbers you can pick (saying the same thing we said about 5 and 3):

If we multiply the sum of any two numbers with the difference of the same two numbers, then we get the difference of the square of the first number, and the square of the second number.

Isn't is simpler to say: $(a+b)(a-b) = a^2 - b^2$ for any two numbers (a and b) you pick?

This example illustrates one aspect of algebra: it is a shorthand notation for numerical relationships that are ALWAYS TRUE, regardless of what numbers you choose. The letters 'a', 'b', 'x', 'y', etc. ALWAYS stand for actual numbers, meaning: any number you pick. The formulas using them are like the famous truism: "two plus two always gives us four".

Most often used algebraic formulas and equations:

$a(b+c) = ab + ac$

$(a+b)(c+d) = ac + ad + bc + bd$

$ab = ba$

$(a/b)(c/d) = ac/bd$

(a/b) / (c/d) = (a/b)(d/c) = ad/bc

$a^0 = 1$

$a^n \times a^m = a^{n+m}$

$a^n / a^m = a^{n-m}$

$(a^n)^m = a^{nm}$

$(ab)^n = a^n b^n$

$(a/b)^n = a^n / b^n$

$(a+b)(a-b) = a^2 - b^2$

$(a+b)^2 = a^2 + 2ab + b^2$

$(a-b)^2 = a^2 - 2ab + b^2$

$a^3 + b^3 = (a+b)(a^2 - ab + b^2)$

$a^3 - b^3 = (a-b)(a^2 + ab + b^2)$

$a^{n/m} = \sqrt[m]{a^n} = (\sqrt[m]{a})^n$

$a^{-n} = 1 / a^n$

$(\sqrt[m]{ab}) = (\sqrt[m]{a})(\sqrt[m]{b})$

$(\sqrt[m]{a/b}) = (\sqrt[m]{a}) / (\sqrt[m]{b})$

Solution to the quadratic equation: $ax^2 + bx + c = 0$

$$x = \frac{-b \pm \sqrt{b^2 - 4ac}}{2a}$$

The binominal theorem:

$(a+b)^n = a^n + na^{n-1}b + (n(n-1)/2!\,)\,a^{n-2}b^2 + (n(n-1)(n-2)/3!\,)\,a^{n-3}b^3 + \ldots + b^n$

here the 'k!' notation (called 'factorial'), where k is a positive integer, means: $k! = 1 \times 2 \times 3 \times \ldots \times k$ (e.g. $3! = 1 \times 2 \times 3 = 6$)

Logarithmic function:

In the exponential function $y = x^n$, we have two variables: the base (x) and the exponent (n). If we know the result (y) and one of the other two, we can find the third.

We already know the n-th root of y, by the $x = \sqrt[n]{y}$ operation.

To determine the exponent (n) from the other two, we use the logarithmic function: $n = \log_x y$. In other words $n = \log_x y$ is the exponent that gives us $y = x^n$.

Examples:

$$\log_3 9 = 2 \text{ because } 3^2 = 9$$
$$\log_5 125 = 3 \text{ because } 5^3 = 125$$

Generally:

$b^{(\log_b x)} = x$

Some useful formulas (easy to prove):

$\log (nm) = \log n + \log m$

$\log (n/m) = \log n - \log m$

$\log (n^m) = m \log n$

$\ln x = \log_e x$ where 'e' is the base of the natural logarithm:

$$e = 1/1! + 1/2! + 1/3! + 1/4! + 1/5! + \ldots$$

converging to the value of 2.71828182 for the first 8 decimals.

Trigonometry

Using a right triangle, we define the trigonometric functions as:

$\sin \alpha = $ opposite side / hypotenuse
$\cos \alpha = $ adjacent side / hypotenuse
$\tan \alpha = \sin \alpha / \cos \alpha$
$\cot \alpha = \cos \alpha / \sin \alpha$

$\sin^2 \alpha + \cos^2 \alpha = 1$

$\sin(\alpha + \beta) = \sin\alpha\ \cos\beta + \cos\alpha\ \sin\beta$
$\sin(\alpha - \beta) = \sin\alpha\ \cos\beta - \cos\alpha\ \sin\beta$
$\cos(\alpha + \beta) = \cos\alpha\ \cos\beta - \sin\alpha\ \sin\beta$
$\cos(\alpha - \beta) = \cos\alpha\ \cos\beta + \sin\alpha\ \sin\beta$

$\sin 2\alpha = 2\sin\alpha\ \cos\alpha$

Differential Calculus

Single variable Functions

In mathematics, a function describes the relationship between two variables. The relationship is denoted as $y = f(x)$, that is the value of the 'y' variable depends on (and is uniquely determined by) the value of the 'x' variable. Examples for this is plenty, such as the distance covered by a free-falling object in Earth's gravitational field: $d(t) = 4.9t^2$ meters if the time is measured in seconds. The distance fallen is a function of time.

Derivation of Single variable Functions

In order to understand Newton's Laws and Theory of Universal Gravity, I have to introduce a mathematical notation. This will save a lot of time later and make the essence of these theories very clear.

Differential calculus is taught in many high-school math programs but, just in case you missed it, here it comes.

We will need it in order to understand simple concepts such as speed and acceleration.

Average speed is easier to understand. We take the total distance traveled and divide it by the total time it took us. Suppose we get 30 miles per hour. This does not mean that we never exceeded the speed limit, because our "momentary speed" during that interval could have reached 100 miles per hour, for a short time, before we slowed down. The cop, measuring our momentary speed with his radar device would have been right in giving us a ticket. So how do we define "momentary speed"?

Get the average speed for smaller and smaller time intervals – so small in fact that there would be no time at all during the interval to speed up or slow down.

The mathematical notation used for the value of speed obtained in this way is the following:

Suppose we use the Greek letter delta: Δ to denote a very small amount. Then Δ s could mean a very small distance traveled and Δ t could mean a very small time interval it took us to travel Δ s distance. Then the ratio Δ s/Δ t will be our average speed for that Δ t time interval.

Now, as we make Δ t smaller and smaller, the ratio Δ s/Δ t will be closer and closer to our true momentary speed.

When Δ t reaches the smallest possible value without being zero (in infinity), then we will denote it with the 'dt' symbol, and the value of Δ s at that point will be denoted by the 'ds' symbol. The 'ds' and 'dt' values are called '**differentials**' of 's' and 't', meaning an infinitesimally small amount of each.

Dividing 'ds' by 'dt' will give us the true momentary speed and it is denoted mathematically as $\frac{ds}{dt}$ which is verbally called the "**derivative** of s with respect to t" and the process of finding this derivative is called: "**differentiating**".

This process of differentiating can be used for any value (let's denote it as 'A'), the change of which is determined by the change of another value (let's denote it as 'B'). We say that the dependent value ('A') is a *function* of the other value ('B'), or mathematically: A=f(B).

So the expression $\dfrac{dA}{dB}$ will mean the value of the A/B ratio

as B gets smaller and smaller ad infinitum.

For example, the **acceleration** of a moving body at any point in time means the rate of change of speed (and or direction) of the moving body at that time.

Let's assume for simplicity's sake that the direction does not change (the car is accelerating on a straight road). Similarly to the definition of momentary speed, we define the momentary acceleration as the a = dv/dt derivative (where 'v' is the speed and 't' is the time).

If 'A' is a complicated expression, then instead of the $\dfrac{dA}{dB}$

notation, I will use the 'd/dB (A)' notation which means exactly the same. For example, the momentum of a moving body is the product of its mass and speed: mv, and the rate of change of the momentum will be denoted as d/dt(mv).

In short: when you see the symbol of d/dt – it always means the momentary rate of change of one variable, as determined by the other variable at that moment.

So, for example, a = dv(t)/dt means the acceleration at that moment, which is the rate of change of speed at a moment of 't'.

Derivation formulas:

Changes of most physical variables are described by some mathematical function. As we have seen, the free-falling body accelerates exponentially, the electric and magnetic fields in space vary by the trigonometric function, the entropy of a closed system changes by a logarithmic function. Therefore, if we want to describe the rate of change for these variables, we need to derive the derivation formulas for these functions.

Without proof (very simple to prove) here are the formulas, together with some useful rules:

$d(c)/dx = 0$ if 'c' is a constant

$d/dx(cf) = c\ df/dx$ if 'c' is a constant

$d/dx\ (f + g) = df/dx + dg/dx$

$d/dx\ (fg) = f\ dg/dx + g\ (df/dx)$

$d/dx\ (f/g) = (f\ dg/dx - g\ (df/dx))\ /\ g^2$

$d/dx\ (y(u(x)) = (dy/dy)(du/dx)$ /chain rule/

$d/dx\ (x^n) = nx^{n-1}$

$d/dx(\sin\ \alpha) = \cos\ \alpha$

$d/dx(\cos\ \alpha) = -\sin\ \alpha$

$d/dx(\ln\ x) = 1/x$

$d/dx(e^x) = e^x$

Multi-variable Functions

In mathematics, a multi-variable function describes the relationship between one variable, depending on the value of 2 or more other variables. The relationship is denoted as z = f(x,y…), that is the value of the 'z' variable depends on (and uniquely determined by) the values of the 'x', 'y',… variables. Examples for this are plenty, such as the value of the pressure of an ideal gas, depending on the values of pressure and temperature: $p = f(V, T) = CT/V$

Derivation of multi-variable functions

The multi-variable function f(x,y,z) can be differentiated by any one of its variables by treating the other 2 variables as constant. We use a different notation for this: $\frac{\partial f}{\partial x}$

For example, if f(x,y,z) = $3x^3 + 2y^2 + 3 z$

$\partial f / \partial x = 9x^2$ (the other two terms are treated as constants)

In the same way:

$\partial f / \partial y = 4y$

$\partial f / \partial z = 3$

Integration

Indefinite Integral:

It is also called the Antiderivative because it is the inverse function of the derivative function.

For example: $d/dx (x^2 + C) = 2x$, (C is a constant) therefore its inverse function' denoted by the symbol of: $\int$ is:

$$\int 2x\ dx = x^2 + C \text{ (where C is any constant)}$$

In the general case: $\int x^n\ dx = (x^{n+1}/(n+1)) + C$

because $d/dx(x^{n+1}/(n+1)) = x^n$

Another example: since $d/dx (\ln x) = 1/x$

$$\int \frac{1}{x} dx = \ln x + C \text{ (where C is any constant)}$$

The indefinite integral $F(x) = \int f(x)dx$ is also a function of x.

Definite Integral:

I have to define another mathematical concept, also invented by Newton and Leibnitz, called: 'integrating'. Imagine that a car keeps accelerating over a period of time. During this time it constantly changes its speed and let's assume that we know its "momentary speed" at any time during this interval.

We want to know the total distance travelled from beginning to the end of this time.

This is how we proceed: Chop up the time interval into 'n' very small Δ t sections. The times at the middle of these intervals are denoted by t_1, t_2, ..., t_i,..., t_n. During each Δ t time the speed is very close to the average $v(t_i)$ where t_i is the middle of that Δ t. The distance traveled during each Δ t period is

$$\Delta s = v(t_i) * \Delta t$$

where t_i is the middle of that Δ t.

If we make Δ t very, very small, than the distance traveled during our whole acceleration period is very close to the sum of all the little distances traveled. We use the $\sum$ symbol to denote this sum:

$$s = \sum_{i=1}^{n} v(ti) * \Delta t$$

Again, just as we did with differentiating if we make Δ t smaller and smaller, ad infinitum, the $\sum_{i=1}^{n} v(ti) * \Delta t$ sum becomes a definite value that we will denote with the "Integral" symbol of: $\int$

$$s = \int_{t1}^{tn} v(t)dt$$

where t_1 is the beginning of the acceleration period and t_n is the end of the same and $v(t)$ means the speed at the exact time 't'.

To find the value of a definite integral of a function f(x) between two values of 'a' and 'b' we use the indefinite integral of the function of f(x) in the following way:

$$\int_a^b f(x)\mathrm{d}x = F(b) - F(a) \text{ where } F(x) = \int f(x)dx$$

For example, because $F(x) = \int x^2 dx = x^3 / 3$

$$\int_2^5 x^2\, dx = (5^3 / 3) - (2^3 / 3) = (5^3 - 2^3)/3) = (125 - 8)/3 = 39$$

The value of a definite integral is always a constant.

In short, when you see a symbol like $\int_a^b f(x)\mathrm{d}x$ it always means the summation of the f(x) values over an interval (a,b) for the value of x.

For example: $d = \int_0^5 v(t)dt$ means the covered distance 'd' (in kilometers) between 0 and 5 hours, if the momentary speed of the vehicle is represented as v(t) km/hour at the time 't' hour.

Vectors

The rest of this math toolkit will be needed only in the Electrodynamics section of the "Next Level" chapter. You can skip it if you only want to have a taste of doing 'real physics' without getting into more advanced mathematical topics.

Vectors, in mathematics and physics, represent values that have both a magnitude and a direction, such as the speed of a moving vehicle, for example.

Vectors in this book are denoted with a bold font, such as: **V**

Vectors in physics have three coordinates, as they represent the three projections of the vector on the three axes of the Cartesian co-ordinte system. Usually denoted as:

$$\mathbf{V}(x,y,z)$$

or $\mathbf{V}(x_1, x_2, x_3)$

The length of the vector $\mathbf{V}$ is denoted by: abs($\mathbf{V}$), meaning the absolute value.

Operations on vectors include addition and subtraction and two different multiplications: the **dot product** and the **cross product**, which are used extensively in Physics.

Let's deal with the following two vectors:

$$\mathbf{X}(x_1,x_2,x_3) \text{ and } \mathbf{Y}(y_1,y_2,y_3)$$

Adding two vectors means adding their respective coordinates:

$$\mathbf{Z} = \mathbf{X+Y}(x_1+y_1,\ x_2+y_2,\ x_3+y_3)$$

The **<u>dot product</u>** of two vectors is a number, according to the following:

$$\mathbf{A\cdot B} = \text{abs}(\mathbf{A})\text{abs}(\mathbf{B})\cos \alpha$$

where α is the angle between the two vectors.

This is equal to the following (easy to prove):

$$\mathbf{A\cdot B} = x_1y_1 + x_2y_2 + x_3y_3$$

Obviously, the dot product of 2 perpendicular vectors is zero, because cos 90 = 0

The **cross product** of two vectors is a vector itself.

It is represented as **Z** = **X** x **Y**

The magnitude of this vector is:

$$abs(\mathbf{Z}) = abs(\mathbf{X})abs(\mathbf{Y})\sin \alpha$$

where α is the angle between the two vectors.

Obviously, the cross product of 2 parallel vectors is zero, because sin 0 = 0

    The orientation of this vector is perpendicular to the plane containing **X** and **Y,** in the direction given by the right-hand rule: point the right-hand thumb in the direction of **X,** the right-hand index finger in the direction of **Y,** and then the right-hand middle finger points in the direction of **X** x **Y.**

    Suppose the first vector represents a vector in the direction of the flow of electricity in a wire, and the second vector represents a vector from the wire to the perimeter of a circle around the wire, then we can see how the cross product vector is always tangential to the circle, anywhere on the circle, and that represents the curling magnetic field around the wire. That is why this cross product is so useful to describe the structure of the field and that is why it is called the 'curl' in Physics.

    If we do the calculation of the coordinates of the cross product vector, from the definition, we will find that the components of the vector are:

$$Z = X \times Y(x_2y_3 - x_3y_2, \; x_3y_1 - x_1y_3, \; x_1y_2 - x_2y_1)$$

The cross product and the dot product are used in Physics to describe the structure of a vector field mathematically.

Scalar Field

In Physics a **field** is a set of values that are a function of space and time coordinates: that is a value that changes from one location to another and changes with time. For example, the temperature of the atmosphere around our planet is described by a scalar field: $T=f(x,y,x,t)$ where T is the temperature, t is the time, and x,y, and z are the space coordinates.

A useful way to describe the rate of change of this value is a mathematical notation called the **gradient of the field** at different points in space. This gradient can be represented by a vector with the following components:

$$\text{grad t } (\frac{\partial f}{\partial x} \; ; \frac{\partial f}{\partial y} \; ; \; \frac{\partial f}{\partial z})$$

obviously this gradient gives us another field with the gradient vector changing with space and time coordinates: that is a vector field.

Vector Field

In physics, we usually deal with what we call a **vector field** which assigns a vector to every point in space, so it becomes a function $V(P)$ where $V(v_1,v_2,v_3)$ is a function of the $P(x_1,x_2,x_3)$ point in space. A good example is the gravitational force field around a planet which defines the strength and direction of the gravitational force (a vector) in each point in space (another vector).

Two other concepts relating to vectors need to be defined: the **divergence** and the **curl** of a vector field. If a vector field has a positive **divergence** at a point in space (often called a 'source' or 'faucet'), it means that the vectors point outward from that location in space, like the repulsive force of a positive charge against all other positive charges. If it has a negative divergence (often called a 'sink' or 'drain'), then the vectors point inward to that location, like the gravitational attractive force of masses.

On the other hand, if a force field has a **curl** (sometimes called a 'whirlpool') at a point in space, then the vectors curl around that location, like the magnetic forces around a current-carrying wire.

Mathematically, the divergence of a vector field **V**(x,y,z) is described by a scalar of the following value:

$$\text{div } \mathbf{V} = = \frac{\partial v1}{\partial x} + \frac{\partial v2}{\partial y} + \frac{\partial v3}{\partial z}$$

where v1, v2, and v3 are the components of the vector **V** and x,y and z are the space coordinates **V** is a function of.

Mathematically, curl of a vector field **V**(x,y,z) is described by a vector field of the components:

$$\mathbf{Z} = \text{curl } \mathbf{V} \ \left(\frac{\partial v2}{\partial z} - \frac{\partial v3}{\partial y}; \ \frac{\partial v3}{\partial x} - \frac{\partial v1}{\partial z}; \ \frac{\partial v1}{\partial y} - \frac{\partial v2}{\partial x} \right)$$

The Del Operator: ∇

At this point, I have to introduce a mathematical notation that doesn't seem to make any sense but will prove to be very useful as a shorthand description of mathematical (and physical) relationships.

We define a new 'vector' that is not a vector in the usual sense because its components are not numbers but mathematical operations. I will call it the "Del" vector and denote it with the following symbol: ∇

Its 'components' are as follows:

$$\nabla \; (\frac{\partial}{\partial x}; \; \frac{\partial}{\partial y}; \; \frac{\partial}{\partial z})$$

By itself the del 'vector' has no meaning – only when we use it in mathematical notations does it have meaning, as we will see in a minute:

With this notation, the expressions associated with gradient, divergence, and curl become considerably simpler and easier to remember.

When we multiply this 'vector' with a scalar field function $T(x,y,z)$, we will get the gradient of the scalar field, as we defined it before:

$$\boxed{\text{grad } T(x,y,z) = \nabla \; T}$$

that is a vector with the components of $\frac{\partial T}{\partial x}$, $\frac{\partial T}{\partial y}$, $\frac{\partial T}{\partial z}$
which is the definition of the gradient, as we described it before.

When we create the dot product between the del operator vector and a vector field $\mathbf{V}(x,y,z)$ where v1, v2, and v3 are the components of the vector **V** and x,y and z are the space coordinates **V** is a function of, we get:

$$\text{div } \mathbf{V} = \nabla \cdot \mathbf{V}$$

and, according to the dot product definition, this will be equal to

$$\frac{\partial v1}{\partial x} + \frac{\partial v2}{\partial y} + \frac{\partial v3}{\partial z}$$

which is the definition of the divergence, as we described it before.

And, finally, when we create the cross product between the del operator vector and a vector field **V**(x,y,z) where v1, v2, and v3 are the components of the vector **V** and x,y and z are the space coordinates **V** is a function of, we get:

$$\text{curl } \mathbf{V} = \nabla \times \mathbf{V}$$

and, according to the cross product definition, this will be equal to a vector with the following components:

$$\left(\frac{\partial v2}{\partial z} - \frac{\partial v3}{\partial y} \;;\; \frac{\partial v3}{\partial x} - \frac{\partial v1}{\partial z} \;;\; \frac{\partial v1}{\partial y} - \frac{\partial v2}{\partial x}\right)$$

which is the definition of the curl, as we described it before.

Finally, some useful theorems that we often use, particularly in Electrodynamics:

The Gaussian integral:

$$\iiint_V (\nabla \cdot \mathbf{F})\, dV = \oiint_S \mathbf{F} \cdot d\mathbf{S}$$

Stokes's theorem:

$$\int_S \nabla \times \mathbf{F} \cdot d\mathbf{S} = \oint_C \mathbf{F} \cdot d\mathbf{r}$$

And one more useful formula we use when deriving the electromagnetic wave function:

$$\nabla \times (\nabla \times \mathbf{E}) = \nabla(\nabla \mathbf{E}) - \nabla^2 \mathbf{E}$$

The Greek alphabet:

Physicists need a lot of letters and symbols for their equations. Here is the table of the Greek letters used in physics:

A	α	alpha	N	ν	nu
B	β	beta	Ξ	ξ	ksi
Γ	γ	gamma	O	o	omicron
Δ	δ	delta	Π	π	pi
E	ε	epsilon	P	ρ	rho
Z	ζ	zeta	Σ	σς	sigma
H	η	eta	T	τ	tau
Θ	θ	theta	Y	υ	upsilon
I	ι	iota	Φ	φ	phi
K	κ	kappa	X	χ	chi
Λ	λ	lambda	Ψ	ψ	psi
M	μ	mu	Ω	ω	omega

<u>The Next Level</u>

In the first part of this book I discussed the human side of science and Physics: I introduced the key scientists and recounted the stories of important discoveries.

However, one more step is required if you really want to understand Physics. In the "Science and Religion" chapter I said that we don't have to take science on faith. Once the basic assumptions are accepted, then we can draw logical conclusions from them and discover new and new principles and applications.

That is what this second part of the book is about: intended for those who feel comfortable with advanced high-school math and are interested to see how the different laws were derived from basic assumptions. They can also follow the detailed solutions to the examples I mentioned earlier.

Even if you are not up on your high school math, the previous chapter: "Your Math Toolkit" explains and summarizes all the mathematical formulas and equations we are using from here on. If you had the confidence to take a look there, you may have found that math is easy and obvious. Then you continue here and feel empowered to do 'real' Physics.

<u>The Science of Newtonian Mechanics</u>

<u>Conservation of Momentum</u>

<u>First case</u>: two bodies interacting, no external forces. Then only internal forces are present that, according to Newton's Law III, they cancel each other out (action-reaction):

$$F_{12} = - F_{21}$$

The total force acting on the system of 2 bodies is

$$F = F_{12} + F_{21} = 0$$

According to Newton $F = ma = m\, dv/dt = d(mv)/dt = dp/dt$ where $p = mv$ momentum.

The total momentum of the system is $p = p_1 + p_2$
therefore
$$F = d/dt(p_1 + p_2) = 0$$

which means that the total momentum of the two objects do not change due to their interaction if no external forces are present.

Second case: multiple-bodies system, possible external forces acting on each body.

Extending Newton's second law to multiple-bodies system (sum only the external forces because internal forces cancel out):

$$d/dt\left(\sum p_i \right) = \sum F_i$$

If we denote the (vector) sum of all external forces with
$$F_{ext} = \sum F_i$$
and the (vector) sum of all the momentums of the individual bodies as $P = \sum p_i$
$$F_{ext} = dP/dt$$

The total external force applied to the system equals the rate of change of the system's total momentum.

If there are no external forces, then the total momentum of the system is conserved ($dP/dt = 0$).

Definition: **Center of mass** of a system of m_i masses is the following vector:

$$R = \sum_{1}^{n} m_i r_i / M$$

Where M is the total Mass of the masses. And since according to Newton's second Law:

$$F_i = d^2 (m_i r_i) / dt^2$$

then

$$\sum_{1}^{n} F_i = F_{ext} = d^2 (\sum_{1}^{n} m_i r_i) / dt^2$$

where F_{ext} is the sum of all external forces (the internal forces cancel out due to Newton's Third Law), therefore:

by using the $\sum_{1}^{n} m_i r_i = RM$ from above,

$$F_{ext} = d^2 (MR) / dt^2$$

or

$$F_{ext} = M d^2 R / dt^2$$

The center of mass moves as if the total mass were concentrated in it and all the forces acted on it.

Applications example: Flight of a rocket expelling fuel at speed 'u' (relative to rocket)

The following table describes the rocket's status at time 't' and a little later, at time 't + Δt'. The observer is sitting in an inertial system outside the rocket.

at time 't'	at time 't + Δt'
before burning Δm fuel **the rocket** mass (including fuel) : m + Δm speed: v **total momentum**: P₁ = (m + Δm)v	**after burning Δm fuel** **the rocket** mass: m speed: v + Δv **the expelled fuel** mass: Δm speed: v-u (opposite direction) **total momentum**: P₂ = m(v + Δv) + Δm(v-u)

According to the conservation of momentum law (where F is the gravitational force acting on the system, as the only external force):

$$\Delta P/\Delta t = (P_2 - P_1) / \Delta t = F$$

and because

$$(P_2 - P_1) = m(v + \Delta v) + \Delta m(v-u) - (m + \Delta m)v = m\Delta v - u\Delta m$$

therefore

$$(m\Delta v - u\Delta m) / \Delta t = F$$

so

$$m\Delta v / \Delta t = F + u\Delta m/ \Delta t$$

or, in differentials,

$$m\, dv/dt = u\, dm/dt + F$$

therefore

$$dv/dt = (u/m)\, dm/dt + F/m$$

and since the gravitational force F = -m g, the change of speed at time t is (negative sign due to Force being opposite to acceleration)

$$\frac{dv}{dt} = \frac{u}{m}\frac{dm}{dt} - g$$

If the initial speed was zero, then speed at time t is:

$$\mathbf{v} = \int (dv/dt)dt$$

and substituting for $\frac{dv}{dt}$ the value obtained above of $\frac{u}{m}\frac{dm}{dt}$ - g

and performing the integral function (according to the rules of calculus), if the rocket starts with zero speed ($t_0 = 0$ and $v(t_0)=0$) then

$$v(t) = u \ln(\frac{m(0)}{m(t)}) - gt$$

which means that after burning all the fuel of mass 'M', carried by the rocket of empty mass 'm', the max speed obtained is:

$$v_{max} = u \ln (\frac{M+m}{m}) - gt$$

which depends only on the mass of the rocket, the amount of fuel carried by the rocket, and the fuel expulsion speed.

The maximum expulsion speed these days with the chemical burning process is about 5000m/sec, but for various losses (heat, etc) only 50% effective speed is realistic. Since the logarithmic function gives a rapidly decreasing value, the fuel requirement grows exponentially.

To increase speed from zero to fuel ejection speed 'u', (M+m)/m has to be 2.718 (the total weight at takeoff has to be 2.7 times of the empty weight of the rocket).

To increase from zero to '2u' (M+m)/m has to be 7.4 (the total weight at takeoff has to be 7.4 times of the empty weight of the rocket).

To increase from zero to '3u' the factor is 20.1
To increase from zero to '4u' the factor is 54.5
That is one reason why multiple-stage rockets are used in spaceships because it sheds useless weight (fuel container, rocket engine, etc) and increases efficiency.

Kinetic Energy, Work, Kinetic Energy Theorem

Let's look at what Newton's Second Law tells us about a body with mass 'm', accelerated by a constant force 'F', for a time interval of 't', increasing its speed uniformly from 'v_1' to 'v_2'. We will use 'x' to denote the distance traveled during this time.

Let's start with Newton's Second Law:
$$F = ma$$
multiplying both sides by x
$$Fx = max$$

The distance traveled during this time is $x = (v_1 + v_2) t / 2$ because the body travels at an average speed of ½ $(v_1 + v_2)$ for 't' time, so

$$Fx = ma (v_1 + v_2) t / 2$$
and since $a = (v_2 - v_1)/t$
$$Fx = [m (v_2 - v_1)/t] (v_1 + v_2) t / 2$$

$$Fx = m (v_2{}^2 - v_1{}^2) / 2$$

$$Fx = m v_2{}^2/2 - m v_1{}^2/2$$

Now, if we don't restrict the situation to constant force and constant acceleration, then we will have the exact same result:

$$F = m \, dv/dt$$

$$\int_{x1}^{x2} F dx = m \int_{x1}^{x2} \frac{dv}{dt} dx = \int_{v1}^{v2} \frac{dx}{dt} dv = \int_{v1}^{v2} v\,dv = m\, \tfrac{1}{2}(v_2{}^2 - v_1{}^2)$$

$$= \tfrac{1}{2}\, m\, v_2{}^2 - \tfrac{1}{2}\, m\, v_1{}^2$$

and if we allow movement along any arbitrary 3-dimensional curve, then assuming that the angle between the force and the dr segment of the curve is: ϑ

$$\int_{r1}^{r2} F dr \cos\vartheta = \tfrac{1}{2}\, m\, v_2{}^2 - \tfrac{1}{2}\, m\, v_1{}^2$$

and if we use vector notation **F** for force and **dr** for curve segment, then

$$\int_{r1}^{r2} \mathbf{F \cdot dr} = \tfrac{1}{2}\, m\, v_2{}^2 - \tfrac{1}{2}\, m\, v_1{}^2$$

where **F·dr** means the dot product of two vectors: **F** and **dr**

Now we will define two new concepts:

Definition: **Work** performed by force F on an object along an arbitrary path is:

$$W = \int_{r1}^{r2} \mathbf{F \cdot dr}$$

Definition: **Kinetic Energy** of a material body of mass 'm', moving at a speed: 'v':

$$T = \tfrac{1}{2}\, m\, v^2$$

Using these definitions, our last equation above can be expressed as:

$$W = T2 - T1$$

which states **that the change in the kinetic energy of the object equals the work performed on it by the force**. "There is no such thing as a free lunch"

The unit of both Work and Energy in the MKS system is 1 Joule = 1 Newton x meter and in the CGS system it is
1 erg = 1 dyn x cm = 10^{-7} Joule

Yet one more new concept is the **Power,** which is the work performed in unit time

$$power = \frac{dW}{dt}$$

with a MKS unit of 1 Watt = 1 Joule/sec

Applications example

A rocket is accelerated from the surface of Earth, vertically up to a velocity v_0 when the engine runs out of fuel and stops. At the point where the engine stopped, the rocket is at height 'R' (measured from the center of Earth), its mass is 'm' and its velocity is 'v_0'. From that point on the rocket's speed is steadily decreasing (the rocket is decelerating) due to the gravitational force acting on it.

What is the escape velocity for the rocket (how large should v_0 be so the rocket does not fall back to Earth)?

According to the Kinetic Energy Theorem, the kinetic energy change (loss) between the time the rocket's speed was v_0 and the time when the rocket decelerated to speed v (while traveling from 'R' to 'r' distance from the centre of Earth)

$$T2 - T1 = W$$

or

$$\tfrac{1}{2} m v^2 - \tfrac{1}{2} m v_0^2 = \int_{R}^{r} \mathbf{F \cdot dr}$$

The gravitational force acting on the rocket at time 't' (with all fuel gone):

$$F(t) = - GMm/r^2$$

where 'M' is the mass of Earth, 'G' is the gravitational constant, 'm' is the unfuelled mass of the rocket and r is its distance from the center of Earth at time 't'. Consequently,

$$\tfrac{1}{2} m v^2 - \tfrac{1}{2} m v_0^2 = \int_{R}^{r} \mathbf{F \cdot dr} = - GMm \int_{R}^{r} 1/r^2 dr = GMm \left(\frac{1}{r} - \frac{1}{R} \right)$$

The rocket reaches maximum 'r' distance when v=0, so

$$-\tfrac{1}{2} m v_0^2 = GMm \left(\frac{1}{r} - \frac{1}{R} \right)$$

$$v_0^2 = 2GM \left(\frac{1}{R} - \frac{1}{r} \right)$$

and since the gravitational force at height 'R' is mg=GmM/ R^2 then G=g R^2/M

$$v_0^2 = 2g (R - R^2/r) = 2gR - 2gR^2/r$$

thus

$$2gR^2/r = 2gR - v_0^2$$

so

$$r = 2gR^2 / (2gR - v_0^2)$$

and

$$r = R / (1 - v_0^2/2gR)$$

The speed v_0 becomes the escape velocity if the max distance 'r' is infinite, which happens when

$$(1 - v_0^2/2gR) = 0 \text{ that is } v_0^2 = 2gR$$

so the escape velocity is independent of the mass of the rocket and its value is:

$$v_0 = \sqrt{2*9.81*6.4} \times 10^3 = 1.1 \times 10^4 \text{ m/sec}$$

or 11km/sec

Conservative forces, Potential Energy

Let's look at the work $\int_{P1}^{P2} \mathbf{F \cdot dr}$ done by the gravitational force

$F = - GMm/r^2$ while moving an object of mass 'm' from point P1 to point P2 in space.

Assuming that the angle between the force and the dr segment of the curve is: ϑ

$$W = \int_{P1}^{P2} \mathbf{F \cdot dr} = \int_{P1}^{P2} F dr \cos\vartheta$$

and since any path from P1 to P2 can be approximated by a jagged path with either radial (ϑ =0) or perpendicular-to-radial (ϑ =90) components, t is obvious that only the radial segments contribute to the integral (cos 0 =1 and cos 90 = 0), so the integral is the same (from distance r_1 to distance r_2), regardless what path we take.
W =

$$W = \int_{r1}^{r2} \mathbf{F \cdot dr}$$

and since F = - GMm/r^2

$$W = - GMm \int_{r1}^{r2} dr/r^2$$

$$W = - GMm \ (1/r_1 - 1/r_2)$$

Definition: A **force field is conservative** if we can move an object from any point P1 to any other point P2 and the work W done depends only on the endpoints and not on the path we chose.

According to this definition, the gravitational force field is conservative.

Definition: The **Potential Energy** U(**P**) of a conservative force field at point **P** is the work done by the force field while moving an object from 'P' to an arbitrarily chosen fixed point P_0 in space. The **potential energy difference** between two points 'P_1' and 'P_2' in a conservative force field will be:

$$U(P_1) - U(P_2) = \int_{P1}^{P0} F \cdot dr - \int_{P2}^{P0} F \cdot dr = \int_{P1}^{P0} F \cdot dr + \int_{P0}^{P2} F \cdot dr = \int_{P1}^{P2} F \cdot dr$$

$$U(P_1) - U(P_2) = \int_{P1}^{P2} F \cdot dr$$

Conservation of Energy

Due to the Kinetic Energy Theorem: the change in Kinetic energy when moving between two points equals the work done against the force: $T2 - T1 = W$

and according to our definition of Potential Energy in a conservative force field, the work done against the force equals the difference in potential energy,

$$U(P_1) - U(P_2) = W$$

Therefore, we can say that in a conservative force field:

$$T2 - T1 = U(P_1) - U(P_2)$$

$$T1 + U(P_1) = T2 + U(P_2)$$

which means that **in a conservative force field the sum of Kinetic and Potential energies remain the same:**

$$T + U = constant$$

That is the total energy of a moving object is conserved.

Potential in a gravitational force field:

As we just determined: the gravitational force field is conservative. The force acting at height 'h' above the surface of Earth is:

$$F = - GMm/(R + h)^2$$

where R is the Earth's radius.

If we select the surface of the Earth for the Potential energy's null point U(0)=0, then the potential energy of the gravitational force field, at height 'h' above the surface:

$$U(h) = U(h) - U(0) = \int_{P1}^{P0} \mathbf{F} \cdot \mathbf{dr} = -GMm \int_{P1}^{P0} 1/r^2 dr =$$

$$GMm \left(\frac{1}{r} - \frac{1}{R} \right)$$

but since r = R+h

$$\left(\frac{1}{r} - \frac{1}{R} \right) = \frac{1}{R+h} - \frac{1}{R} = \frac{R-R-h}{R(R+h)} = - \frac{h}{R(R+h)}$$

and since close to the surface $R \cong R+h$

therefore U(h) = -G Mmh/R^2

and since g=GM/ R^2 then G = g R^2/M

then close to the Earth's surface the potential energy is:

U(h) = -mgh

I could have arrived at the same solution if I used the fact that the gravitational force is the same F=mg close to the Earth's surface, so we would need W = mgh work to move it to height 'h'

If we look at the general case of a small mass 'm' being 'r' distance from the center of a big mass 'M', then assuming the null potential at infinite distance from M (at point P_0), then using our previous equation:

$$U(r) = -GMm \int_{P1}^{P0} 1/r^2 dr = -GMm \left(\frac{1}{r} - \frac{1}{\infty} \right)$$

$$U(r) = -GMm/r$$

Let's recalculate the escape velocity of the object with mass 'm' by using the **Conservation of Energy theorem** (we have done it before by using the Kinetic Energy Theorem):

The total (kinetic + potential) energy close to the Earth's surface is:

$$\tfrac{1}{2} mv^2 - GMm/R$$

The total (kinetic + potential) energy of the rocket at infinity will be zero, because the potential energy there is zero (null point at infinity), and the kinetic energy is also zero because we assumed that the rocket will have decelerated to zero speed at infinity (that is how the escape velocity was defined).

According to the conservation of energy theorem, the total energy (kinetic + potential) remained the same,

consequently:

$$\tfrac{1}{2} mv^2 - GMm/R = 0$$

$$v^2 = 2GM/R$$

and since $g = GM/R^2$ then $G = gR^2/M$, so

$$v^2 = 2gR$$

which is the same value we got before by using the Kinetic Energy Theorem.

Rotating Motion

Rotation is three dimensions

Newton's Second Law states for one point mass:

$$\mathbf{F} = m\mathbf{a} = m \, d^2r/dt^2$$

and for an arbitrary shape and size material body (made up of N pieces of point masses)

$$\sum \mathbf{F_i} = \sum m_i \, d^2r_i/dt^2$$

where the left side is the sum of all external forces (because the internal forces all cancel out due to Newton's Third Law)

and applying the vector product with r_i to both sides, we get (after we exchange the two sides):

$$\sum [\, r_i \times \mathbf{F_i}] = \sum m_i [\, r_i \times d^2r_i/dt^2]$$

and since d/dt [r_i x dr/dt] = [r_i x d^2r_i/dt^2] + [dr_i /dt x dr_i /dt] and the second term is zero because it is the cross product of a vector with itself (sin (0) = 0), then [r_i x d^2r_i/dt^2] = d/dt [r_i x dr_i/dt]

so our previous equation can be written as:

$$\frac{d}{dt}\sum m_i \ [r_i \ x \ dr_i/dt] \ = \ \sum \ [\ r_i \ x \ F_i]$$

so

$$\frac{d}{dt}\sum [r_i \ x \ p_i]= \sum \ [\ r_i \ x \ F_i]$$

Now we define the **Angular Momentum relative to a point in space** of the whole body as:

$$L = \sum L_i = \sum [r_i \ x \ p_i]$$

And we define the total **Torque relative to a point in space** acting on the whole body as:

$$T = \sum T_i = \sum [\ r_i \ x \ F_i]$$

Where r_i are position vectors from that point to the N pieces of point masses m_i.

And now we are ready to state the Law of Angular Momentum for arbitrary 3-d bodies as (derived straight from Newton's second and third laws):

$$dL/dt = T$$

The total torque from all external forces acting on a material body equals the rate of change of its total angular momentum.

This implies the same conservation of angular momentum law:

If the torque from all external forces acting on a material body is zero, then the body's angular momentum remains the same constant value.

Our experience shows that just as material bodies have inertia, a resistance to force and acceleration that opposes either speeding up or slowing down, bodies also have another kind of inertia: a resistance to forces trying to speed up or slow down rotation around some axis.

To describe this new kind of inertia, I will introduce a few new concepts and we will see that Newton's Laws are sufficient to describe the relationship among them and thus mathematically describe the rotating motion of material bodies.

<u>Circular rotation of a point-like material body with mass 'm' at distance 'r' from a fixed axis:</u>

<u>Angular velocity</u> is the rate of change of the rotational angle:

$$\omega = \frac{d\theta}{dt}$$

In a circular motion, the length of an arc 'l' on the circumference of a circle with radius 'r' satisfies the following:

$$\theta / 2\pi = l / 2\pi r$$

(if θ is measured in radians, because the full circle's angle is 2π radian and its circumference is $2\pi r$)

so

$$\theta = l / r$$

The tangential speed of any point with 'r' distance from the axis is:

$$v = dl/dt$$

so

$$v = r \, d\theta / dt$$

thus

$$v = r \, \omega$$

Angular acceleration is the rate of change in the angular velocity: $\alpha = \dfrac{d\omega}{dt}$

From experience we know that resistance to rotation around a given axis depends on the mass distribution around the axis (the closer to the axis, the easier to rotate) and the mass of the body (the bigger the mass the harder to rotate)

Moment of inertia (around a given axis)

for an arbitrary 3-dimensional body is defined as $I = \sum m_i r_i^2$ and for a point-like material body with mass 'm' at distance 'r' from a fixed axis it will become:

$$I = m \, r^2$$

Angular momentum of a rotating body (relative to an axis) was defined as

$$L = [r \times p]$$

and for a point-like material body with mass 'm' at distance 'r' from a fixed axis it will become:

$$L = rmv = rm\,(r\,\omega) = m\,r^2\,\omega = I\,\omega$$

$$L = I\,\omega$$

the product of its 'moment of inertia' and its angular speed:

Torque of a rotating body (relative to an axis) was defined as

$$T = [r \times F]$$

and for a point-like material body with mass 'm' at distance 'r' from a fixed axis it will become:

$$T = F\,r$$

(the farther from the axis we push, the easier it is to rotate the bodies)

Also, according to the conservation of angular momentum theorem: $dL/dt = T$

so

$$T = I\,\frac{d\omega}{dt} = I\,\alpha$$

<u>Example of acceleration of solid cylinder of radius R, mass M, rolling down an inclined plane of angle β.</u>

Speed of center of Mass point Q is 'v', acceleration of same point is 'a', angular velocity is ω, frictional force is 'F'.

Since

$$v = \omega R,$$

then

$$a = \alpha R$$

and

$$T = I \alpha$$

also

$$T = R F = I a / R$$

On the other hand:

$$Ma = Mg \sin\beta - F = M g \sin\beta - Ia/R^2$$

from Newton's 2nd Law
so

$$a(M + I/ R^2) = Mg \sin\beta$$

that gives us:

$$a = (Mg \sin\beta) / (M + I/ R^2)$$

multiplied by R^2

$$a = (MR^2 g \sin\beta) / (M R^2 + I)$$

Moment of inertia for solid cylinder is $I = MR^2/2$
so

$$a = (2g \sin\beta) / 3$$

Independent of Mass and Radius of a cylinder.

For a hollow cylinder $I = MR^2$, so $a = (g \sin\beta) / 2$

and it will lose the race with the solid cylinder, every time, regardless of mass and radius.

Kepler's Second Law

As we have stated before in the Law of conservation of angular momentum:

If the sum of all external torques on a material body is zero, then the body's angular momentum remains the same constant value.

If we apply this law to the Planetary orbits, we will get Kepler's Second Law from the Law of conservation of angular momentum.

We choose an axis going through the Sun, perpendicular to the plane of the planet's orbit.

Since the elliptical orbits of the planets are almost circular in our Solar System, we can use our result for the angular momentum we obtained for circular motions above:

$$L = m \, r^2 \, \omega$$

There is no external torque on the system because the force is always toward the axis (the Sun), so there is no tangential component to the force.

No external torque means that the angular momentum 'L' of the planet will have to stay the same.

$$L = m\,r^2\,\omega = \text{constant}$$

It immediately gives us the result that the planet will speed up when it gets closer to the sun and it will slow down as it gets farther (if 'r' gets smaller, then the only way L can stay constant is by increasing ω and vice versa).

Isn't it marvelous, how a massive object as the planet 'knows' about this law and as it gets close to the Sun, it 'wants' to obey the law and the only way it can do that is by compensating via speeding up in a precise mathematical manner?

The same law explains how ice skaters do pirouettes (as they pull their arms closer to their bodies, the rotation speed has to increase to keep the angular momentum constant).

The same law explains how divers control the rotational speed of their bodies (as they bend their bodies more, the rotation speed has to increase to keep the angular momentum constant).

The Science of Electrodynamics

By the end of the nineteenth century, all crucial experiments were performed, all the basic variables defined and the most important laws formulated. Electrodynamics was ready to become a precise, practical, and consistent science, ready to be used by scientists and engineers to build new technology.

Just like the Science of Mechanics was waiting to be developed after Newton's synthesis, so was the Science of Electrodynamics in need of a systematic summary in the form of a self-contained theory.

There was one stumbling block that needed to be overcome: it had to be built on a precise and consistent definition of the basic variables, and the known formulas had to be expressed in terms of these variables.

Once this task was done, the confusing mass of experimental results and competing theories could be distilled into a surprisingly few fundamental variables and laws.

The basic formulas accepted by all scientists at the time Maxwell finished his work were:

	Electricity Magnetism	Induced Magnetism	Force on moving charge
Coulomb	$F=k_e\, q_1q_2/r^2$ $F=k_m\, p_1p_2/r^2$		
Oersted		Discovery	
Biot-Savart		$df = k_B\, I\, ds\, \sin \zeta\, / r^2$	
Ampére		$F = k_A\, I_1\, I_2\, L/r$	
Lorentz			$F=k_L Qv \sin \zeta$

What remained to be done was to establish some convention, both practical and reasonable, for the proportionality factors k_e, k_m, k_B, k_A, and k_L used in these formulas.

Units and factors

The five different proportionality factors: k_e, k_m, k_B, k_A and k_L are not all independent from each other. We can easily deduce formulas connecting some of them.

Determining the k_e, k_m, k_B relationship:

It was understood that the three parameters of k_e, k_m and k_B can not be independent of each other. Some kind of relationship is obvious if we think it over: If we select k_e and k_m arbitrarily, that will fix the units of electric charge and magnetic pole strength. With that, we will have fixed the unit of current as well, due to the definition of current as charge divided by time.

This means that we can not select the value of k_B arbitrarily, because it would also define the unit of current (from the Biot Savart law), independently from how we have already done it by setting k_e and k_m. The question, as before, was the exact nature of the relationship between these three values.

The first experiment to determine this relationship was performed by Weber and Kohlrausch in 1856. They used a mathematically   deduced   consequence   of   the

$$df = k_B I ds \sin \zeta / r^2$$

Biot-Savart's Law, if applied to a circular current, since the formula for the force on a unit magnetic pole in the center of the circle with radius 'r', as deduced from the Biot Savart Law, is:

$$F = k_B \, 2\pi \, I \, / \, r$$

Weber and Kohlrausch set up an experiment in which they repeatedly (at known frequency 'f') charged a condenser (with known capacitance 'C') with a battery (of known tension 'U') and then discharged it through a circular current with a magnetic needle of unit strength in the center.

They could determine the current going through the galvanometer by calculating the charge per second flowing through the condenser (from 'C', 'U' and 'f') as $I = fCU$.

Finally, in possession of the value of 'I' and 'r', they could determine the value of k_B from the $F = k_B \, 2\pi \, I \, / \, r$ formula by measuring the force on the unit-strength magnetic pole in the center of the circular current.

To their surprise, they found that, if they arbitrarily chose $k_e = 1$ and $k_m = 1$, (thereby choosing the Gaussian units for the electric charge and the magnetic pole) then **the value of k_B turned out to be 1/c** where 'c' is the well-known speed of light of approximately 300,000 km/sec. The exact relationship among k_e, k_m and k_B is as follows: $\dfrac{\sqrt{k_e \; k_m}}{kB} = c$ ($c \cong 3 \times 10^{10}$ cm/sec)

Determining the k_e / k_A ratio:

The first experiment to determine this relationship was performed by Maxwell in 1868.

Let's look at two of the fundamental laws of electrodynamics:

The first of these laws is the:

$$F = k_e\, q_1 q_2 / r^2$$

Coulomb's Law for electric forces. Coulomb did not have a unit for electric charges, he only knew that half as strong charges (easily created by touching the charged body with an identical uncharged one) resulted in a quarter the force, etc. So he knew that the force of attraction (or repulsion) was proportional to the amount of the charges.

However, once we select a certain amount of electric charge as the unit, then the value of k_e can be determined by measuring the force between two unit charges at unit distance from each other. On the other hand, if we arbitrarily select a value for k_e, then it will determine the unit of electric charge: it will be equal to the charge which produces a force equal to the value of k_1, on an identical charge at unit distance from it.

The second of these laws is the:

$$F = k_A\, I_1\ I_2\ L/r$$

Ampére's law for the force of attraction (or repulsion) between two long and parallel wires of length 'L', at distance 'r' ($r \ll L$) from each other, carrying currents I_1 and I_2.

It is obvious that if we select one particular strength of the electric current as a unit current, then the value of k_A will be determined by measuring the force between two parallel wires of unit length, at unit distance from each other, both carrying this unit current.

It is also obvious that k_e and k_A are not independent of each other, due to the $I = Q/t$ definition of electric current: if the unit charge is determined by selecting a value for k_e, then it also determines the unit of electric current and thus the value of k_A. On the other hand, if we choose a value for k_A, then the value of k_e can be calculated from the same formulas.

Maxwell's experiment: performed in London in the spring of 1868, tried to balance the electrostatic attraction between two charged metal plates against the magnetic repulsion between two current-carrying coils. The results of the experiment gave a value about 7% below the value obtained by Weber and Kohlrausch (which was about 4% higher than the actual speed of light determined later more precisely by Foucault).

The relationship between k_e and k_A was found to be:

$$k_e / k_A = c^2/2$$

Determining the k_A, k_m, k_B relationship:

From the Weber-Kohlrausch ($\dfrac{\sqrt{ke\ km}}{kB} = c$) experiment and the Maxwell experiment ($k_e / k_A = c^2/2$), we can easily see that the relationship between k_m, k_A and k_B is:

$$k_A = 2\ k_B{}^2 / k_m$$

For historical reasons, not one, but actually, two different conventions were chosen (similar to the metric and imperial measuring systems) to fix the values of the five proportionality factors. Both are self-consistent and it is easy to convert from one to the other. Some scientists and engineers prefer to work in one, some in the other, depending on their philosophical 'bend' and the particular application they are working on. These two unit systems are the CGS and the MKSA systems.

The CGS System has not been used much in recent years, so I will describe it in the "Next Level" section of the book, together with conversion values to MKSA. *In the rest of the book I will be using the MKSA System exclusively.*

In the MKSA system we will consider the Electric Current one of the fundamental units and deduce all other values from the following four fundamental units: Length (**M**eter), Mass (**K**ilogram), Time (**S**econd) and Electric Current (**A**mpere).

The fourth basic unit of Ampere is defined by arbitrarily selecting (for historical reasons):

$$k_A = 2 \times 10^{-7}$$

We have two more arbitrary choices to make and we will select

$$k_B = 1/4\pi$$

$$k_L = 1$$

By these choices the units we end up with will be practical, considering the typical currents, potentials, etc. used in our technology.

Due to the $k_A = 2 k_B^2 / k_m$ relationship we can calculate the value of k_m as:

$$k_m = 10^7 / (4\pi)^2$$

and from the $k_e / k_A = c^2/2$ relationship we can calculate the value of k_e as:

$$k_e = c^2 \times 10^{-7}$$

where 'c' is the speed of light ($c = 2.9979 \times 10^8$ m/sec), so $k_e \cong 9 \times 10^9$.

So, the formulas, expressed in the MKSA System will be as follows:

	Electricity Magnetism	Currents	Induced Magnetism
Coulomb	$F = (1/4\pi\varepsilon_0) q_1 q_2/r^2$ $F = 10^7 /(4\pi)^2 \, p_1 p_2/r^2$		
Volta		$I = Q/t$	
Oersted			discovery
Biot-Savart			$df = (1/4\pi) \, I \, ds \, \sin \zeta / r^2$
Ampére			$F = 2 \times 10^{-7} \, I_1 \, I_2 \, L/r$

The $\varepsilon_0 \cong 1/(4\pi \times 9 \times 10^9) \cong 8.85 \times 10^{-12}$ constant is called the "**permittivity of free space**". The name 'permittivity' is misleading because it is derived from the study of the behavior of electric fields inside matter, thus suggesting that vacuum is a special kind of matter, instead of the 'lack of matter' that it actually is. However, for historical reasons, the name is stuck, so we might as well accept it without attributing any special meaning to it. Using this value, the proportionality factor:

$$1/4\pi\varepsilon_0 = (1/4\pi)(1/\varepsilon_0) = (1/4\pi)(4\pi \times 9 \times 10^9) = 9 \times 10^9$$

just as we obtained it above for k_e.

One may question why the value 'π' was used. The reason for this is the fact that measuring the force acting on a unit magnetic pole in the center of a circular current (from Biot-Savart law) is: $F = k_B\, 2\pi\, I/r$ (where 'I' is the current and 'r' is the radius of the circular conductor and the circumference of the circle is $2\pi r$).

The reason that the value of 'c' was used is due to the Weber-Kohlrausch experiment that introduced this value into electrodynamics, so it will have to be included in one of the three coefficients.

Summarizing the values of the electrodynamic proportionality factors gives us the following table:

k_e	k_m	k_L	k_B	k_A
$c^2 \times 10^{-7} = k_e = 1/4\pi\varepsilon_0$	$\dfrac{10^7}{(4\pi)^2}$	1	$1/4\pi$	2×10^{-7}

Electric and Magnetic Quantities

Now that the basic formulas are defined in the two different unit systems (after we got rid of the previously unknown proportionality factors), we can define the fundamental variables used in electrodynamic calculations. In Physics, the definition of any fundamental variable is meaningless without precise measuring instructions, and I am going to provide those, for each of the variables, in the MKSA unit systems.

The basic variables used in electromagnetic calculations are the following:

Electric charge (Q)

Electric field in vacuum (E)

Electric Potential (U)

Electric Capacitance (C)

Electric current (I)

Resistance (R)

Magnetic pole strength (p)

Magnetic field in vacuum (B)

Electric charge (Q)

Measuring electric charge can be done by torsion balance, invented by Charles Coulomb.

In the MKSA system the unit of electric charge is called 'Coulomb' and the definition of this unit is: 1 Coulomb electric charge is the charge that repulses another identical charge, in a vacuum, at the distance of one meter, with 9×10^9 Newton force.

Electric field in vacuum (E)

According to experience, the cumulative effect of an electric charge distribution (any number of electric charges distributed in space) exerts a force on any other electric charge moved into any point in space. We measure this force by moving a test charge of value 'Q' to the point 'P' and measure the force acting on it. Therefore, we can assign a value to any point 'P' in space, which will be the force exerted on this charge 'Q' divided by 'Q' (to get the force on unit charge) at that point. We denote this value by the letter E(P) and call it the value of the electric field at point 'P'.

In the MKSA system, the unit of the electric field has no special name and the definition of this unit is: 1 MKSA unit of the electric field is the force that repulses a 1 Coulomb test charge, in a vacuum, at the distance of one meter, with a 1 Newton force.

Electric potential (U)

If we recall our discussion in Part I about the potential in a gravitational field, then we can very simply understand the concept of electrical potential as well, because the electrostatic attraction follows the same rule as the gravitational attraction: the field of a point charge is spherical and inversely proportional with the square of the distance. For this reason, the mechanical work required to move a unit charge from point 'A' to point 'B' does not depend on the path we choose.

Therefore, each point in space can have a number assigned to it, which is the mechanical work required to move a unit charge to that point, from an arbitrarily selected 'null-point' (by convention we use the surface of Earth for this). This value assigned to point 'P' will be called the potential of the electric field at that point. The difference between the electrical potential between two points in space is often referred to as the electrical "tension" and we can measure this value in a number of ways, the simplest of which is a simple electroscope.

The same reasoning applies to any electric field, due to any charge distribution, because of the superposition principle, the total field can be summed up (integrated) as the sum of the individual field due to point-like charges making up the distribution.

The power of electric batteries is usually specified by giving the electric tension between the two poles (positive and negative) of the battery. We will see (below) that for a given load on the battery (resistance of the circuit connected to it) the current flowing through the circuit will depend on the tension of the battery: the bigger the tension is, the bigger the current will be.

In the MKSA system, the unit of electric potential is called 1 Volt (V) and the definition of this unit is: a point 'P' in an electric field has a potential of 1 Volt, if we need to perform 1 Joule mechanical work to move a 1 Coulomb charge from ground to point 'P'. The potential difference (or electric tension) between points 'A' and 'B' is 1 Volt, if we need to perform 1 Joule mechanical work to move a 1 Coulomb charge from point 'A' to point 'B'. 1 Volt = 1 Joule/Coulomb

Electric Capacitance

If we select two electric conductors of arbitrary shape and size and we give one of them a charge of +Q, while giving the other a charge of −Q, then we will have the same electric tension 'U' between any two points, one on each of the two pieces of conductors. This is obvious from the fact that the electric field is always perpendicular to the surface of a conductor when the system is in equilibrium (otherwise we would have a current) so we don't have to perform any work by moving a charge from any one point to any other on the same surface.

If we connect the poles of a battery (of tension U) to the 2 conductors, then the conductors will be charged: one to a certain value +Q and the other to −Q. The amount of the charge Q will depend on the **geometrical properties of the two conductors and their mutual positions.**

If we double the tension of the battery (from U to 2U) on both conductors, then the charge Q will also double (from 'Q' to 2Q) because we will have to perform twice as much work (the electric field is proportional with the charges) moving a unit charge from one conductor to the other. So the charge Q is proportionate to the tension between the conductors:

$$Q = CU$$

This proportionality factor 'C' is called the **Capacitance** of the two-conductor system (called 'condenser'). The Capacitance of the condenser can be measured by measuring the electric tension between the two surfaces of the condenser and the electric charge on any of the surfaces.

In the MKSA system, the unit of electric Capacitance is called 1 Farad, and the definition of this unit is: the capacity of a condenser is 1 Farad, if we have a 1 Volt of tension between the two conductors of the condenser when we gave them a +1 and – 1 Coulomb charge. By the C=Q/U definition

1 Farad = 1 Coulomb/1 Volt = 1 As/Volt (Ampere x Second/Volt)

Electric current (I)

As we have seen, Voltaic batteries create a continuous flow of electric charges along conducting wires. This is called an electric current and is measured by measuring the charge that crosses over a section of wire in a unit of time. This can be accomplished by measuring the tension in the circuit and a capacity of a condenser that gets charged by this current. This is how currents were measured before the current's magnetic effect was discovered and the relationship between current and magnetic field was established. After that galvanometers were used to measure electric currents.

In the MKSA system, the unit of electric current is called 1 Ampere and the definition of this unit is: 1 Ampere of electric current is the current that carries 1 Coulomb charge over the cross-section of the wire in one second.

This definition for the Ampére is conceptually clear, but it is very difficult to measure the amount of electric charge moving across a wire in a unit time, so a more practical definition is being used in real life. In experiments, it is easier to measure the force between two parallel wires, so a practical definition of the MKSA unit of electric current is based on Ampére's law for the force acting between two current-carrying wires:

An electric current is 1 Ampere if, flowing through two very long parallel wires one meter apart, in a vacuum, it generates 2 x 10^{-7} Newton force per each 1 meter long section.

Resistance (R)

George Simon Ohm (1787 – 1854), a German physicist, experimented with wires of different lengths and diameters and found a simple relationship between the tension of the battery and the current in the wire, depending on the characteristics of the wire making up the circuit. He found that the current was inversely proportional to the length and directly proportional to the area of the cross-section of the wire. Thus he established that the current-carrying capacity of the wire depended only on the material used and the geometrical properties, and was independent of the tension in the circuit. We call this simple relationship Ohm's Law and it states that

$$I = U/R$$

where 'I' is the current, 'U' is the tension and 'R' is the resistance of the conductor. The formula says that the electric current is directly proportional to the electric tension and inversely proportional to the resistance of the conductor.

In the MKSA system the unit of electric resistance is called 1 Ohm and the definition of this unit is: 1 Ohm of resistance results in 1 Ampere of current if we have 1 Volt of tension at the two ends of the wire.

Magnetic pole (p)

If we magnetize two very thin and very long steel needles, we find that the ends either attract or repel each other. We assume that the two ends of the needle are magnetic poles with a pole strength of 'p' and the attractive or repulsive force according to Coulomb's measurement is directly proportional with the pole strengths and inversely proportional with the square of their distance:

$$F = k_m \, p_1 \, p_2 \, / \, r^2$$

(we can create different strength poles by putting two, three, etc identical magnetic needles side by side). We can measure the force by Coulomb's torsion balance. The unit of magnetic pole strength in the two different systems are as follows:
In the MKSA system, the unit of the magnetic pole is called 'Weber' and the definition of this unit is: 1 Weber of the magnetic pole is the pole that repulses another identical pole, in a vacuum, at the distance of one meter, with $10^7 / (4\pi)^2$ Newton force.

The magnetic field in vacuum (B)

Before Oersted's discovery, the induction of magnetic field wasn't known yet, so the magnetic field was defined similarly to the electric field, as follows: the cumulative effect of a magnetic pole-distribution (any number of magnetic poles distributed in space) exerts a force on any other magnetic pole moved into any point in space. We measure this force by moving a test magnetic pole of value 'p' to the point 'P' and measure the force acting on it. Therefore, we can assign a value to any point 'P' in space, which will be the force exerted on this magnetic pole 'p' divided by 'p' (to get the force on unit magnetic pole) at that point. We denote this value by the letter H(P) and call it the value of the magnetic field at point 'P'.

However, after Oersted's discovery and Biot-Savart's and Ampére's force laws were known, the definition and measurement of the magnetic field changed to a more reliable one. The original definition, based on magnetic poles was cumbersome, mostly because no true magnetic pole existed (only dipoles). Now we use the magnetic field created by a current-carrying wire to define the field and measure its strength.

Now we can use the Lorentz force formula to define a magnetic field, without using magnetic dipoles: as described before, on a point-like electric charge 'Q', moving at speed 'v' in a magnetic field, the field exerts a force, the magnitude of which is equal to:

$$F = k_L \, B \, W \, v \, \sin\vartheta$$

The new value of 'B', representing the strength of the magnet (exerting a force on the moving charge) will be, by definition, the magnetic field surrounding the magnet.

By choosing a value for the k_L factor arbitrarily, the unit for the magnetic field will have been determined as well.

In the MKSA system the unit of a magnetic field is called 1 Tesla (T) and the definition of this unit is: 1 Tesla of a magnetic field is the field that exerts 1 Newton force, in a vacuum, on a charge of 1 Coulomb, moving at a speed of 1 m/sec, perpendicular to the field. 1 Tesla = 1 Vs/m^2

Induced magnetic field

According to the Lorentz force formula, the force acting on charge Q moving with speed 'v' in a magnetic field 'B' is:

$$F = B \, Q \, v \, \sin\vartheta$$

Let's assume that the direction of the magnetic field is perpendicular to the d rection of 'v', so $\sin\vartheta = 1$.

If we take a wire of length 'L', carrying a current of 'I$_1$' then since a charge 'Q' moving along length 'L' of wire with the speed of 'v' constitutes a current cf

$$I_1 = Q \, v \, / L$$

The reason for this is the following (we will reason in the CGS system).

If we have a total charge of Q Franklins on a wire of length L cm, then on 1 cm length of the wire we have a charge of Q/L Franklins.

Since the charge distribution moves at v cm/sec along the wire, then the length of the charge distribution moving across a cross-section of the wire in one second is v cm.

Therefore the total charge moving across the cross-section of the wire in one second is   vQ/L Franklins.

Therefore:
$$I_1 = Q \, v \, / L$$
so
$$Q \, v = I_1 \, L$$

and substituting this into the Lorentz-force Law, the force acting on the wire carrying a current 'I_1' in a wire of length 'L' will be

$$F = B \, I_1 \, L$$

On the other hand, if the magnetic field 'B' is produced by another parallel wire, carrying a current of 'I_2', 'd' distance from the first one, then, according to Ampére, the force acting between the two parallel wires is:

$$F = (2 \times 10^{-7}) \, I_1 \, I_2 \, L \, / \, r$$

Equating the last two formulas on 'F', we get the following equation for B:

$$B = (2 \times 10^{-7}) \, (I_2 \, / \, r)$$

This gives us the value of a magnetic field at distance 'r' from a straight, long wire, carrying an electric current 'I_2'.

This result was obtained by applying the Lorentz Force formula and Ampere's Force Law for a long, straight, current-carrying wire.

Now, if we start out from the Biot Savart Law:

$$df = (1/4\pi) \, I \, ds \, \sin \zeta \, / \, r^2$$

This is the **<u>force</u>** a wire segment 'ds' carrying a current 'I' exerts on a unit magnetic pole 'r' distance from the wire.

It is useful to rewrite (without detailed proof) the Biot Savart Law to express the **magnetic induction** vector at point P, at distance 'r' from the current-carrying wire:

$$B(P) = (1/4\pi\varepsilon_0 \; c^2) \; I \int \; dl \; x \; r/r^2$$

It can also be expressed as:

$$B(P) = (\mu_0/4\pi) \; I \int \; dl \; x \; r/r^2$$

where the $\mu_0 = 4\pi \; x \; 10^{-7}$ constant is called the "**permeability of free space**".

This name is derived from the study of the magnetic field inside matter, which is outside the scope of this book, so we just have to accept the term without worrying too much about the meaning attached to the name.

In both cases, the coefficients $(1/4\pi\varepsilon_0 \; c^2) = (\mu_0/4\pi) = 10^{-7}$

If we now integrate along an infinite wire (from an angle -$\pi/2$ to $+\pi/2$), then we get back the same value for B as in the formula we deduced above from the Lorentz Force formula and Ampere's Force Law:

$$B = (2 \; x \; 10^{-7}) \; (I \; / \; r)$$

That is the value of a magnetic field at distance 'r' from a straight, long wire, carrying an electric current 'I'.

<u>Induced electric field</u>

Let us consider a closed electric circuit.

When we have an electric current through a conducting wire, we know that electric charges are moving through this wire from the negative to the positive terminal of the battery.

Obviously, there is a force acting on these charges that make them move. Part of this force is due to the electric charges accumulated on the terminals of the battery.

Another part of the force is the electrostatic force along the length of the wire, due to the accumulation of charges coming out of the battery and building up in wire segments. This electromagnetic force 'wants' to smooth out the charge distribution along the wire, so the moving charges will get uniformly distributed, once an equilibrium is reached.

We define the **electromotive force (emf)** as "the tangential force per unit charge in the wire integrated over length, once around the complete circuit."

$$\varepsilon = \oint_L E dl$$

where E is the electric field vector (force per unit charge) and we integrate it over the whole circuit.

Since we considered only the tangential component of the force (perpendicular component does not contribute to the current) on a unit charge and we integrated over the whole wire (segments of which may experience varying forces), we can see that the **emf** uniquely describes the external conditions (whatever they are) that cause the electric current. Emf's can be generated in a wire in three different ways:

1.   By moving the wire near a magnet

2.   By moving a magnet near a wire
3.   By changing the current in a nearby wire"

The obvious question: what is common in these three cases?

The answer is of two parts:

1.   in all three cases, we have a vector field present that we have already called the "magnetic field"

2.   In all three cases, the magnetic field changes **relative** to the wire containing the electric charges.

Is there a simple way to express the generated emf as the function of this relative change of the magnetic field?

Yes, there is.

In vector analysis, we defined the **flux of a vector field** through a surface as the integral of the perpendicular (to the surface) component of the vector field over the entire surface.
Using this definition, we can now define the **magnetic flux** passing through the surface defined by the electric wire.

$$\Phi = \oint_S Bds$$

If we consider a special case of a rectangular circuit with length 'l' and width 'w', as it is pulled out of a perpendicular and uniform magnetic field, with speed 'v' to decrease the magnetic flux through the loop, the flux through the loop is:

$$\Phi = \oint_S Bds = -Blw$$

as long as the whole loop is in the magnetic field.

If we start pulling the loop out of the magnetic field, then the flux will decrease at the rate of:

$$\delta\Phi/\delta t = -Bw \; dl/dt = -vBw$$

where 'v is the speed of the moving loop.

On the other hand, a unit electric charge in the wire segment moving in the magnetic field with speed 'v', experiences an **E= v x B** Lorentz Force that will be equal (because B and v are perpendicular) to E = vB.

Due to the definition of the electromotive force:

$$\varepsilon = \oint_L E dl = \oint_L vB \, dl = vBw$$

because the 'lengthwise' segments of the circuit contribute nothing to the emf because there the tangential component of the force is zero (the **E= v x B** Lorentz Force in those segments is perpendicular to the wire, so no tangential component exists). The other segment with width 'w' is already out of the magnetic field, so that segment does not contribute to the emf either (B=0 there).

Combining the last two equations, finally, we have what we call the **Flux Rule:**

$$\varepsilon = -\delta\Phi/\delta t$$

that is **the Electromotive Force in a closed circuit is equal to the rate of change of the flux through the circuits**.

The beauty of the flux rule is in its generality: according to many experiments performed in all scenarios, it applies equally in all three cases considered above (regardless of how the magnetic flux is generated). It also applies to a circuit of any arbitrary shape, moving in an arbitrary direction through a non-uniform magnetic field. In fact, the loop need not even maintain a fixed shape.

If we recall that we defined the potential difference (or **electric tension**) between points 'A' and 'B' as the mechanical work we need to perform to move a unit charge from point 'A' to point 'B' in an electric field E:

$$V = \oint_L Edl$$

and Ohm's Law states that the current in the wire equals:

$$I = V/R$$

therefore

$$I = (1/R) \oint_L Edl = \varepsilon \ 1/R$$

Clearly, the current flowing through a closed electric wire is proportional to the electromotive force in the closed circuit.

Self-inductance, Lenz's Law

Suppose we connect a circuit to a battery by closing a switch. The electric current needs some time to build up from zero to the value determined by Ohm's law. During that time we have a changing electric current that produces a changing magnetic field surrounding the wire. Since the wire is immersed in the same changing magnetic field, the magnetic flux going through the circuit is changing as well.

According to our Flux Rule, an induced emf is generated in the wire, due to the changing flux, due to the changing electric current.

The effect of this induced emf is summarized by Lenz's Law:

"the [induced] emf tries to oppose any flux change [that gave rise to it]. That is, the direction of an induced emf is always such that if a current were to flow in the direction of the emf, it would produce a flux of B that opposes the change in B that produced the emf. ... It tries to keep the current constant; it is opposite to the current when the current is increasing, and it is in the direction of the current when it is decreasing"

In a sense the electric current is trying to cut the branch it is sitting on (oppose the only thing that its existence depends on). Looks like it is a general rule of nature, not only a typical human attitude we are so familiar with.

The same induced emf is observed when we open the switch, disconnecting the circuit from the battery. The suddenly decreasing current (on its way down to zero) produces a suddenly changing magnetic field which means a suddenly dropping flux, so it generates an induced emf that wants to keep the current going. That is why we observe a spark in the switch sometimes when opening t.

Maxwell Equations

The four Maxwell Equations we are familiar with today were not completely new Physics, unknown before Maxwell formulated them. Unlike Kepler, whose laws were brand new, never before suspected by anyone, Maxwell synthesized the crucial discoveries out of the thousands of pieces of knowledge physicists accumulated during the past centuries.

It was Maxwell who identified the minimum number of fundamental facts that would explain everything else. He also put them into a modern mathematical form of vector differential equations, using Faraday's field concept that no one had taken seriously before.

Once Maxwell completed his equations, other physicists could deduce logical consequences from them and predict other yet unobserved phenomena that also ought to be true, if Maxwell's equations correctly described nature.

The only addition to known facts by Maxwell himself was the extra term he suggested to be used in his fourth equation to complete the symmetry of the electromagnetic system. This extra term is the one that led to the discovery of electromagnetic radiation that is the foundation of our entire communication technology today.

Maxwell had two equations for the electric field (divergence and curl) and two equations for the magnetic field (divergence and curl).

In the following descriptions of Maxwell's laws, I will be using the MKSA unit system as explained before.

The first equation (Gauss's Law)

Two experimental laws were built into Maxwell's first equation. They are the following:

- Coulomb's law
- Superposition Principle

These two facts can be expressed as the equation describing the Electric field as a function of space coordinates as follows: The electric field (electrostatic force on unit charge) at an arbitrary point P_0 will be equal to:

$$E(P_0) = 1/4\pi\varepsilon_0 \int_V \rho(P)dV/r^2$$

where
- $\rho(P) = \lim \Delta Q/\Delta V$ (as ΔV approaches zero) is the **charge distribution** (charge per unit volume at point P) [in units of Coulomb per cubic meter]

- 'r' is the distance of ΔV from P_0. [in units of meters]

- ε_0 is called the "**permittivity of free space**" and its value is:
$\varepsilon_0 = 8.85 \times 10^{-12}$ [in units of C^2/Nm^2]

We can define the **Flux of the Electric force field** through any surface as the integral of the perpendicular (to the surface) component of the vector field over the entire surface.

$$\Phi = \int_S Eds$$

Let's consider the flux of E for a point charge Q, through a sphere of radius 'r':

The value of E(P), according to Coulomb's Law is E(P) = $1/4\pi\varepsilon_0$ Q/r^2

So the value of the Electric Flux, due to a point charge Q is:

$$\Phi = \int_S Eds = (1/4\pi\varepsilon_0)\,(Q/r^2) \oint_S ds$$

and the integral $\oint_S ds$ over a sphere of radius 'r' is the $4\pi r^2$

surface area of the sphere, therefore:

$$\Phi = Q/\varepsilon_0$$

as we can see, the flux is independent of the radius of the sphere!

This value will remain the same if we replace the sphere with any arbitrary closed surface that surrounds the charge, as it was demonstrated by Richard Feynman in his "Lectures on Physics" II-4-8.

Finally, if we use an arbitrary charge distribution ρ(P), enclosed within the sphere, due to the superposition principle, the flux will be equal:

$$\Phi = \text{(the sum of all charges inside S)} /\varepsilon_0$$

that is

$$\Phi = \oint_S Eds = 1/\varepsilon_0 \int_V \rho(P)dV$$

This is the Integral form of Maxwell's First Equation, which can also be expressed in differential form, by using the Vector Analytical Gauss Law, according to which:

$$\oint_S Eds = \int_V (\nabla \cdot \mathbf{E}) \, dV$$

where the $\nabla \cdot \mathbf{E}$ is a vector dot product.

From the last two equation we get:

$$\int_V (\nabla \cdot \mathbf{E}) \, dV = 1/\varepsilon_0 \int_V \rho(P) dV$$

and since the integral is true for any Volume, then we get the differential form of Maxwell's First Equation:

$$\nabla \cdot \mathbf{E}(P) = \rho(P)/\varepsilon_0$$

or, using a different notation:

$$\text{div } \mathbf{E}(P) = \rho(P)/\varepsilon_0$$

states that the divergence of the Electric field surrounding a point equals the electric charge density at that point divided by ε_0 (the source of the electric field is always electric charge).

The second equation (Faraday's Law)

The physical content of Maxwell's second equation contains Faraday's Flux rule in differential form. When discussing magnetic induction, I defined the magnetic flux through a surface 'S' as:

$$\Phi = \int_S Bds$$

I also stated the Flux Rule as:

$$\mathcal{E} = -\,\delta\Phi/\delta t$$

or, using the definitions for $\mathcal{E}$ (the emf in the circuit) and Φ

$$\oint_L E dl = -\,\delta/\delta t \oint_S B ds$$

This is the Integral form of Maxwell's Second Equation, which can also be expressed in differential form, by using the Vector Calculus Stokes theorem, according to which:

$$\oint_L E dl = \oint_S (rotE)ds$$

we get the differential form of Maxwell's Second Equation:

$$\nabla \times \mathbf{E} = -\,\delta\mathbf{B}/\delta t$$

or, using a different notation:

$$rot\ \mathbf{E} = -\,\delta\mathbf{B}/\delta t$$

states that the curl of an electric field around a point equals the rate of change of the magnetic field at that point (Faraday's induction experiment).

The third equation

Remember that the magnetic field was defined via the Lorentz Force: if there is a force acting on a moving charge even if there is no measurable Electrical field present, then it is due to the magnetic field $\mathbf{B}(P)$ and the force equals $\mathbf{F}=Q(\mathbf{v} \times \mathbf{B})$.

Since a charge Q, moving with speed 'v' represents an electric current, we can also use the Biot-Savart Law to describe the magnetic field produced by a current 'I' flowing through a conductor as (in the MKSA unit system):

$$\mathbf{B(P)} = (1/4\pi\varepsilon_0 c^2)\, I \int \mathbf{dl} \times \mathbf{r}/r^3$$

For the general case of volume current, where the flow of charge is distributed throughout a three-dimensional region, we have to define a current density as follows: Consider a "tube" of infinitesimal cross-section 'da', running parallel to the flow. If the current in this tube is 'dI', the **volume current density** is:

$$\mathbf{J} = \frac{dI}{da}$$

that is the current per unit area-perpendicular-to-the-flow. Let's rewrite the Biot-Savart Law for volume currents, by using this definition, as:

$$\mathbf{B(P)} = (1/4\pi\varepsilon_0 c^2) \int_V (\mathbf{J} \times \mathbf{r}/r^3)dV$$

It can be shown that taking the divergence of this vector field, we will have:

$$\nabla \cdot \mathbf{B} = 0$$

or in a different notation:

$$\text{div } \mathbf{B} = 0$$

which is the differential form of Maxwell's Third Equation.

From vector Calculus we know that:

$$\int_V (\nabla \cdot \mathbf{B})dV = \oint_S Bds$$

and if we use our definition for magnetic flux as:

$$\Phi = \oint_S Bds$$

then we have the integral form of Maxwell's Third Equation as:

$$\Phi = \oint_S Bds = 0$$

stating that the flux of the magnetic field through any closed surface is zero (due to no magnetic monopole ever observed).

The fourth equation (Ampére's law)

The Biot-Savart Law for volume currents is:

$$B(P) = (1/4\pi\varepsilon_0 c^2) \int_V (J \times r/r^3)dV$$

Applying the curl operator to this vector field:

$$\nabla \times B = (1/4\pi\varepsilon_0 c^2) \int_V \nabla \times (J \times r/r^3)dV$$

and, applying the $A \times (B \times C) = B(AC)-(BA)C$ product rule from vector analysis, we get:

$$\nabla \times B = J/\varepsilon_0 c^2$$

Maxwell had a problem with this equation when he got this far in his calculations. Even though it was derived from the Biot-Savart Law which is known to be correct, however, mathematically this equation will lead to contradiction if we multiply both sides by the '∇' del operator:

$$\nabla (\nabla \times B) = (1/\varepsilon_0 c^2) \nabla J$$

The left side is always zero (we know it from vector analysis)

The right side can be rewritten as $-(1/\varepsilon_0 c^2)\, d\rho/dt$, but $d\rho/dt$ can not be zero because we can move charges around inside the volume and that would change the charge density in time. Maxwell realized that this mathematical difficulty could be solved if he arbitrarily added another term to the right side of the $\nabla \times \mathbf{B} = \mathbf{J}/\varepsilon_0 c^2$ equation, making it look like:

$$\nabla \times \mathbf{B} = \mathbf{J}/\varepsilon_0 c^2 + (1/c^2)\, dE/dt$$

which is the correct and final Fourth Maxwell Equation in differential form, stating that the curl of a magnetic field at a point equals the sum of the current density at that point (divided by $\varepsilon_0 c^2$) and the rate of change of the electric field at that point (divided by c^2).

Maxwell's suggestion for adding the rate-of-change-of-the-electric-field term into the right side of the fourth equation is a very appealing idea, not just mathematically but physically as well: it suggests a fundamental symmetry: we know that the change of electric field results in an induced magnetic field (second equation), it would be reasonable to expect that the reverse should be true as well: any change in the magnetic field ought to induce an electric field. Maxwell was absolutely right about this suggestion, as it was experimentally verified later.

These four equations, together with the 'Lorentz Force Law' and the conservation of electric charge assumption covers absolutely everything known then about electrodynamics.

With these equations, if we know the charge density and current density in space, as a function of space and time, then, with the help of the Maxwell equations, we can find out the value of the Electric and Magnetic fields as well, as a function of space and time.

Electromagnetic waves

One consequence of the modified Fourth Equation is the conclusion that electromagnetic fields travel as a transverse wave (perpendicular to the direction of travel) at the speed of light. Let's consider the Maxwell Equations in a region of space without any charges and/or currents. Then the four equations will look like this:

I./ $\nabla \cdot \mathbf{E} = 0$

II./ $\nabla \times \mathbf{E} = -d\mathbf{B}/dt$

III./ $\nabla \cdot \mathbf{B} = 0$

IV./ $\nabla \times \mathbf{B} = (1/c^2) \, c\mathbf{E}/dt$

In II. and IV the variables **E** and **B** are mixed together which makes it difficult to solve the differential equations. However, we can un-mix them mathematically by applying the curl operation to both equations:

from vector analysis we know that:
$$\nabla \times (\nabla \times \mathbf{E}) = \nabla(\nabla\mathbf{E}) - \nabla^2\mathbf{E}$$
which is, because of I./
$$\text{(i)} \ \nabla \times (\nabla \times \mathbf{E}) = -\nabla^2\mathbf{E}$$
on the other hand, because of II./
$$\nabla \times (\nabla \times \mathbf{E}) = \nabla \times (-d\mathbf{B}/dt) = -d/dt \, (\nabla \times \mathbf{B})$$
and because of IV./
$$\text{(ii)} \ \nabla \times (\nabla \times \mathbf{E}) = -(1/c^2)d^2 \, \mathbf{E}/dt^2$$

and combining (i) with (ii) we get:
$$\text{(a)} \ \nabla^2\mathbf{E} = (1/c^2)d^2 \, \mathbf{E}/dt^2$$
and by similar operations we get:

(aa) $\nabla^2 \mathbf{B} = (1/c^2)d^2 \mathbf{B}/dt^2$

The last two equations are second-degree differential equations and we know from mechanics that these equations describe waves that travel at speed 'c'.

When we look at what is happening *physically*, when we jiggle a charge for a short period of time, we see that the moving charge, constituting an electric current, will

1.   create a changing magnetic field (the field has to build up from zero to its final value) according to the Biot-Savart Law).

2.   But the changing magnetic field induces a changing electric field (the electric field also has to build up from zero to its final value) according to the second Maxwell equation).

3.   This changing electric field will create a changing magnetic field (the magnetic field has to build up from zero to its final value) according to the modified fourth Maxwell equation

And then 2. and 3. keep alternating – forever in vacuum (or until the energy of the fields is absorbed by matter), even if we remove the original charge from the point of origin.

Because of (a) and (aa), it is easy to see that both the electric and magnetic fields will propagate in space, perpendicular to both each other and the direction in which they are going.

The CGS system

In the CGS System, we deduce all other values from the following three fundamental units: Length (**C**entimeter), Mass (**G**ram), and Time (**S**econd).

By agreement, k_e anc k_m were arbitrarily selected to be equal to 1, and thus, according to the above: $\dfrac{\sqrt{ke\ km}}{kB} = c$ formula, k_B must be $k_B = 1/c$.

k_L was also arbitrarily selected as: $k_L = 1/c$.
So the five parameters in the CGS system are as follows:

$$k_e = 1$$
$$k_m = 1$$
$$k_L = 1/c$$
$$k_B = 1/c$$
$$k_A = 2/c^2$$

(due to the $k_A = 2 \times k_B^2 / k_m$ relationship)

One important relationship that we will need later is:
$$k_A/2k_L = k_B$$

So, the formulas, expressed in the CGS System will be as follows:

	Electricity Magnetism	Induced Magnetism	Force on moving charge
Coulomb	$F=q_1q_2/r^2$ $F=p_1p_2/r^2$		
Volta			
Oersted		discovery	
Biot-Savart		$df = (1/c)I\ ds \sin \zeta / r^2$	
Ampére		$F = (2/c^2)\ I_1\ I_2\ L/r$	
Lorentz			$F=(1/c)Qv \sin \zeta$
Faraday			

In the CGS system the unit of electric charge is called 'Franklin' and the definition of this unit is: 1 Franklin electric charge is the charge that repulses another identical charge, in a vacuum, at the distance of one centimeter, with a 1 dyn force. 1 CGS unit = 1 $cm^{3/2}$ $g^{1/2}$ s^{-1} (the dimension is defined by Coulomb's Law)

The conversion factor between Franklin and Coulomb is:

$$1 \text{ Coulomb} = 3 \times 10^9 \text{ Franklin}$$

In the CGS system, the unit of the electric field has no special name and the definition of this unit is: 1 CGS unit of an electric field is the force that repulses a 1 Franklin test charge, in vacuum, at the distance of one centimeter, with a 1 dyn force. 1 CGS unit = 1 $cm^{-1/2}$ $g^{1/2}$ s^{-1} (dimension defined by the E = F/Q formula)

The conversion factor between the two units is:

$$1 \text{ MKSA unit} = 1/(3 \times 10^{4)} \text{ CGS unit}$$

In the CGS system, the unit of electric potential has no special name and the definition of this unit is: a point 'P' in an electric field has a potential of 1 CGS unit, if we need to perform 1 erg mechanical work to move 1 Franklin charge from ground to point 'P'. The potential difference (or electric tension) between points 'A' and 'B' is 1 CGS unit, if we need to perform 1 erg mechanical work to move a 1 Franklin charge from point 'A' to point 'B'. 1 CGS unit = 1 $cm^{1/2}$ $g^{1/2}$ s^{-1}

(dimension defined by the U = Work/Q formula)

The conversion factor between the two units is:

$$1 \text{ Volt} = 1/299.79 \text{ CGS unit}$$

In the CGS system, the unit of electric Capacitance is 1 cm and the definition of this unit is: the capacity of a condenser is 1 cm, if we have a 1 CGS unit of tension between the two conductors of the condenser when we gave them a +1 and −1 Franklin charge. By the C=Q/U definition

1 CGS unit = (1 cm$^{3/2}$ g$^{1/2}$ s^{-1}) / (1 cm$^{1/2}$ g$^{1/2}$ s^{-1}) = 1 cm

The conversion factor between the two units is:

$$1 \text{ Farad} = 8.987 \times 10^{11} \text{ cm}$$

In the CGS system, the unit of electric current has no name and the definition of this unit is: 1 CGS unit of electric current is the current that carries 1 Franklin charge over the cross-section of the wire in one second.

1 CGS unit = (1 cm$^{3/2}$ g$^{1/2}$ s^{-2}) = 1 Fr s^{-1} (dimension defined by the I = Q/time formula). The conversion factor between the two units is: 1 Ampere = 2.9979 x 10^9 CGS unit

In the CGS system, the unit of resistance has no name and the definition of this unit is: 1 CGS unit of resistance results in 1 CGS unit of current if we have 1 CGS unit of tension at the two ends of the wire.

1 CGS unit = 1 cm^{-1} s (dimension defined by the R=U/I formula)

The conversion factor between the two units is:

$$1 \text{ Ohm} = 1/(8.987 \times 10^{11}) \text{ CGS unit}$$

In the CGS system, the unit of the magnetic pole has no name and the definition of this unit is: 1 CGS unit of the magnetic pole is the pole that repulses another identical pole, in a vacuum, at the distance of one centimeter, with a 1 dyn force.

1 CGS unit = 1 cm$^{3/2}$ g$^{1/2}$ s^{-1} (dimension defined by the magnetic Coulomb formula)

The conversion factor between Weber and CGS unit is:

1 Weber = $10^8/4\pi$ = 0.796 x 10^7 CGS unit.

In the CGS system, the unit of the magnetic field is called 1 Gauss (G) and the definition of this unit is: 1 Gauss of the magnetic field is the field that exerts '1/c' dyn force, in a vacuum, on a charge of 1 Franklin, moving at a speed of 1 cm/sec, perpendicular to the field (where c=2.9979 x 10^{10})

1 CGS unit = 1 $cm^{-1/2}$ $g^{1/2}$ s^{-1}

The conversion factor between the two units is:

1 Tesla = 10^4 Gauss

The Science of Thermodynamics

The Ideal Gas Law

In the chapter of Thermodynamics we described the **Combined Gas Law:**

$$pV = mRT$$

where 'p' and 'V' are the pressure and volume of the (mass 'm') gas at temperature 'T' Kelvin degrees and 'R' is a constant that depends only on the kind of gas we are using ($R = p_0V_0/mT_0$)

This experimentally obtained formula can *be* deduced mathematically from Bernoulli's kinetic theory of gases if we make a few further simplifying assumptions:

1.   Consider a cylindrical volume in the vertical position (with a piston on top)
2.   Piston area is 'A' and displacement from the bottom is 'h'
3.   The number of gas molecules in the piston is 'N'
4.   The gas molecules can go only in one of six directions left, right, forward, backward, up, and down - with N/6 molecules moving in each direction.
5.   Each molecule has the same speed 'v' and mass 'μ'

These assumptions will simplify the calculation and, if we consider averages, they are not too far from reality. With these assumptions, we can calculate the following:

The current volume of the gas is:

$$V = Ah$$

Since the depth of the cylinder is 'h', then all the upward moving molecules (N/6 number of them) will collide with the piston within h/v seconds, therefore, the number of molecules colliding with the piston in one second is: N/6 divided by h/v, that is:

$$Nv/6h$$

The total momentum change in each second on the piston is:

$$I = (Nv/6h)\ 2\mu v = N\mu v^2 / 3h$$

According to Newton, the upward force on the piston is: $F = dI/dt$, that is the momentum change in unit time, which is what we calculated above:

$$F = N\mu v^2 / 3h$$

Pressure is defined as force per unit area, therefore:

$$p = F/A = (N\mu v^2 / 3h) / A$$

and since $A = V/h$ (where V is the volume of gas)

$$p = N\mu v^2 / 3V$$

so

$$pV = N\mu v^2 / 3$$

Knowing that the kinetic energy of each molecule is:

$$E_K = \mu v^2 / 2$$

We can rewrite our equations as:

$$pV = (2/3)\ N\ E_K$$

If we compare this equation with the experimentally obtained ideal gas law:

$$pV = mRT$$

then we get the same formula if we assume that the absolute temperature of an ideal gas depends on the number of molecules and their average speed in the volume of gas according to the following formula:

$$T = (2/3)\ (N/mR)\ E_K$$

Now, since 'μ' is the mass of one gas molecule, then its total mass 'm' is $m = N\mu$, therefore:

$$T = (2/3)\ (1/\mu R)\ E_K$$

If we define a new constant, called **Boltzmann's constant** as:

$$k = \mu R$$

then the ideal gas law will become:

$$pV = NkT$$

If we measure the mass of the gas with the new mass-unit (the "mole") , then the **value of 'k' does not depend on the kind of gas we are using**, because:

$$k = \mu R = \mu \, (p_0 V_0 / m T_0)$$

and the value μ /m is the reciprocal of Avogadro's constant, if 'm' is measured in moles (in each 'mole' grams of any gas there is the same $N_A = 6.023 \times 10^{23}$ number of molecules present). Boltzmann's constant then has the same value for any kind of gas and it is:

$$k = 1.38 \times 10^{-16} \, erg/^\circ K$$

Rewriting the $T = (2/3) \, (1/\mu R) \, E_K$ formula for the definition of the temperature (and using the $k = \mu R$ definition of 'k') gets us:

$$E_K = (3/2) \, kT$$

which basically says that the average kinetic energy of a gas molecule, in any volume of gas, is proportional to the temperature of the gas in the volume, and the factor of this proportionality is independent of the chemical composition of the gas (same for any kind of gas).

Rearranging the equation gives us:

$$T = (2/3k) \, E_K$$

Which says that the temperature of any gas is uniquely determined by the kinetic energy of the gas molecules, and it is determined the same way for any kind of gas, that is the temperature does not depend on the chemical composition of the gas molecules (this is, of course, how we defined temperature above, so the theory and experiment would agree with each other).

Energy, the First Law of Thermodynamics

As we have seen in Mechanics, the concept of energy is closely related to the concept of work. It was shown that one of the consequences of Newton's second law is the following:

Definition: **Work** performed by force F on an object along an arbitrary path is:

$$W = \int_{r1}^{r2} \mathbf{F \cdot dr}$$

Definition: **Kinetic Energy** of a material body of mass 'm', moving at a speed: 'v':

$$T = \frac{1}{2} m\, v^2$$

Using these definitions, we can prove that the change in the kinetic energy of an object equals the work performed on it by the force. This is called the *Kinetic Energy Theorem* in the Science of mechanics.

From this, we can see that energy will do work for us. It can lift weights, pull carts, send us to the Moon, wash our laundry. It is a good thing, a useful thing. We don't want to lose it, so we are happy to hear that it is conserved. However, this conservation does not always seem to work. Sometimes it seems that a definite amount of mechanical energy just disappears without a trace.

Isaac Asimov, in his "Asimov's chronology of Science and Discovery" describes the thought process that produced the First Law of Thermodynamics very eloquently:

"…In real life, though, a moving body *does* stop moving after a while because of air resistance or because of friction with the ground. What happens to its energy then? Perhaps it is converted to heat. If so, however, a given amount of mechanical energy should be converted into some fixed amount of heat."

What Joule did was clever and convincing. He knew about electricity producing heat and he wanted to know what the produced heat depended on. After careful experimentation he published his finding in what is known today as the Joule Heating Law:

$$Q_{cal} = K V^2 t / R$$

where 'V' is the potential difference in the wire, 't' is the time the current is on, 'R' is the resistance of the wire and 'K' is a constant: K=0.239 cal/joule.

Next, he measured the amount of mechanical work required (to run the generator) to produce the electricity, to obtain the relationship between mechanical work and heat in the indirect **mechanical work – electricity – heat** transformation.

Following this, he set up his famous paddle-wheel experiment where he could measure the same relationship in the direct **mechanical work – heat** transformation. To his delight, the relationship proved to be the same, which suggested that the mechanical energy was conserved in at least two different transformations to heat.

From this, it was clear that heat is a form of energy and Joule published his experimental result for the conversion factor between mechanical energy and heat energy. This conversion factor is now known as the "Mechanical equivalent of heat" and its value is: 4.18×10^7 erg of mechanical work = 1 calorie (in Joule's honor, a new unit for energy (work) was introduced later:
$$1 \text{ joule} = 10^7 \text{ erg}$$

so the **"mechanical equivalent of heat"** is: 4.18 joule of mechanical work = 1 calorie of heat

Soon after Joule's experiments both Helmholtz and Thomson published his announcement of the conservation of energy principle.

For thermodynamics, it simply means that the total energy content of a system changes only by the heat put into it (positive or negative) and the work done on it (positive or negative). This is called the **First Law of Thermodynamics**:

$$\Delta U = \Delta Q + \Delta W$$

This relationship is considered to be correct for any system, going through any process that involves heat given to, or taken out of, the system.

For the special case of a heat engine that uses some kind of gas that is heated, and due to its heat-related expansion exerts pressure on some moving part of the machine (piston) producing mechanical work, we can reformulate the first law into a more useful form, as follows.

First, let's define pressure as the perpendicular component of force acting on a unit surface:

$$p = \Delta F / \Delta S$$

Experimental data show that the relative volume expansion $\Delta V / V$ is proportional with the pressure:

$$-\Delta V / V = \xi \, p$$

where ξ is called the "compressibility" of the material and it depends on the material used. It can be shown (by calculating the work done while compressing a cube with size 'a' to size a-Δa) that the mechanical work done when compressing a volume of gas by ΔV, at pressure 'p' equals $-p\Delta V$, so the total work done on a volume of gas while compressing it from volume V_1 to V_2 equals:

$$L = -\int_{V1}^{V2} p\,dV$$

which gives us a more useful form of the first law of thermodynamics for the special case of heat engines:

$$\Delta Q = \Delta U + \int_{V1}^{V2} p\,dV$$

If we put ΔQ heat into the heat engine, it will increase the internal energy by ΔU and produces $\int_{V1}^{V2} p\,dV$ useful mechanical work expanding the volume against pressure.

For ideal gas, due to the combined gas law: $pV = CT$ then

$p = CT (1/V)$ so the integral (according to basic calculus), assuming constant T temperature,

$$\Delta Q = \Delta U + CT \ln (V_2 / V_1)$$

The internal energy of a given amount (mass 'm') of ideal gas

Another consequence of the first law of thermodynamics and the ideal gas law is a formula for the internal energy of a given amount (mass 'm') of the ideal gas.

Let's consider the following experiment: If we put two identical cylinders into a calorimeter, one of which contains an ideal gas at high pressure and the other contains a vacuum, then open a valve between the two cylinders to let part of the gas expanding into the second cylinder, the calorimeter shows that no heat transfer took place (the calorimeter's temperature remained same). Nor was any work performed, so the internal energy of the ideal gas remained the same (according to the first law).

This internal energy can not depend only on the pressure of the gas, because the pressure dropped to half due to the expansion, yet the energy remained the same. Due to the same reasoning, this internal energy can not depend only on the volume of the gas, because the volume doubled due to the expansion, yet the energy remained the same. The internal energy can only depend on the product pV (this product remained the same) which, from the ideal gas law, depends only on the temperature ($pV = CT$). Therefore, the internal energy must be a function of the temperature only: $U = f(T)$.

Let's now look at what happens to a mass 'm' of ideal gas if we keep its volume constant ($dV=0$) and put into it a small amount of heat: $dQ > 0$. According to the first law of thermodynamics: $dQ = dU + pdV$, but since $dV=0$, $dQ = dU$.

On the other hand, according to the definition of specific heat:
$c_v = (1/m) \, dQ/dT$

$$dQ = c_v m dT$$

but $dQ = dU$ (see above) so $dU = c_v m dT$, therefore

$$dU/dT = c_v m$$

so

$$c_v = (1/m)\, dU/dT$$

Since according to the $U = f(T)$ result (see above) U depends only on the temperature, then the specific heat c_v can depend only on the temperature. However, according to experimental results, specific heat does not depend on the temperature either. With this in mind, integrating from $dU/dT = c_v m$ gives us:

$$U = mc_v T + U_0$$

Where U_0 is the internal energy of the ideal gas at the $T=0$ absolute zero point. We can choose $U_0 = 0$ because all we ever work with is energy *differences*, where U_0 falls out, therefore:

$$U = mc_v T$$

that is the internal energy of ideal gases is proportionate to the absolute temperature.

Carnot's Cycle and reversible engines

The four phases of the Carnot-cycle are the following:

Isothermal expansion (same temperature) of the gas in the cylinder (against outside pressure), at temperature T_1, increasing its volume from V_A to V_B, while its pressure drops from p_A to p_B. During this expansion, the gas extracts Q_1 heat from the boiler and performs

$$Q_1 = L_1 = -\int_{V_A}^{V_B} pdV \text{ work.}$$

Since according to the combined gas law: $pV = CT$ then

$p = CT (1/V)$ so the integral (according to basic calculus)

$$Q_1 = L_1 = CT_1 \ln (V_B / V_A) > 0 \qquad\qquad (1)$$

Adiabatic expansion (no heat transfer, due to perfect insulation) increases the volume from V_B to V_C while its pressure drops from p_B to p_C and its temperature drops from T_1 to T_2. Since there is no heat transfer, according to the first law of thermodynamics, all the work performed by the gas decreases its internal energy:

$$L_2 = U_2 - U_1 = mc_v(T_2 - T_1) \qquad < 0 \quad (2a)$$

According to the first law, for any small section of the adiabatic expansion:

$$dQ = dU + pdV = 0$$

however,

$$U = mc_vT$$

and

$$p = mCT/V$$

therefore

$$dQ = mc_vdT + mCTdV/V = 0$$

and thus

$$dT/T + c \, dV/V = 0$$

where $c = C/c_v$ is a constant value.

Integrating the last equation we get:

$$\ln T + c \ln V = const$$

therefore

$$\ln (TV^c) = const$$

which gives us:

$$TV^c = const \qquad (2b)$$

Isothermal compression (same temperature) of the gas in the cylinder, at temperature T_2, reduces its volume from V_C to V_D, while its pressure increases from p_C to p_D. During this compression the gas transfers $Q_2 > 0$ heat into the cooler and performs

$$- Q_2 = L_3 = - \int_{V_1}^{V_2} pdV \ \text{work} < 0$$

Since, according to the combined gas law: $pV = CT$ then

$p = CT(1/V)$ so the integral (according to basic calculus)

$$- Q_2 = L_3 = CT_2 \ln (V_D / V_C) < 0 \quad (3)$$

Adiabatic compression (no heat transfer, due to perfect insulation) decreases the volume from V_D to V_A while its pressure increases from p_D to p_A (we return to the starting point of the cycle), and its temperature increases from T_2 to T_1. Since there is no heat transfer, according to the first law of thermodynamics, all the work performed by the gas decreases its internal energy:

$$L_4 = U_1 - U_2 = mc_v(T_1 - T_2) \qquad < 0 \qquad (4)$$

The complete cycle:

The total work performed by the gas will be the sum $L = L_1 + L_2 + L_3 + L_4$

It is easy to see that $L_2 + L_4 = 0$

therefore
$$L = L_1 + L_3$$

$$L = CT_1 \ln (V_E / V_A) \ + \ CT_2 \ln (V_D / V_C) \quad (5)$$

But since for adiabatic processes TV^c = const, we have

$$T_1 V_B{}^c = T_2 V_C{}^c \text{ and } T_1 V_A{}^c = T_2 V_D{}^c$$

and dividing these two ecuations we get:
$$V_B / V_A = V_C / V_D \qquad (6)$$
Therefore from (5)
$$L = CT_1 \ln (V_B / V_A) \ - \ CT_2 \ln (V_C / V_D)$$
therefore
$$L = C(T_1 - T_2) \ln (V_B / V_A)$$
But, according to (1)
$$Q_1 = L_1 = CT_1 \ln (V_B / V_A)$$
from which we get
$$L/Q_1 = (T_1 - T_2) / T_1$$
therefore

$$L/Q_1 = 1 - T_2/T_1 \qquad (7)$$

And this is Carnot's formula for the maximum efficiency (frictionless, reversible engine) that can be obtained from heat engines. It is amazing that Carnot could come up with this result, without knowledge of the first law of thermodynamics that we heavily depended on in our calculations.

The Second Law of Thermodynamics - Entropy

As we have seen in our analysis of the Carnot cycle, according to (3)

$$-Q_2 = L_3 = CT_2 \ln (V_D / V_C) = - CT_2 \ln (V_C / V_D)$$

and because of (6)
$$Q_2 = CT_2 \ln (V_B / V_A)$$
we also get
$$Q_1/Q_2 = T_1/T_2$$
or
$$Q_1/T_1 = Q_2/T_2 \qquad (8)$$

In order to arrive at the mathematical form of the second law, we will have to consider two consequences of the second law as stated in its first form. The two consequences are the following:

1.   If the second law is true, then Carnot's efficiency formula for reversible engines is true for any material used in the heat engine (not just ideal gas)

2. If the second law is true, and if any section of the Carnot cycle is irreversible, then the efficiency of the whole machine has to be less than $(T - T_2) / T_1$

<u>To show that the first consequence is true</u>, let's imagine that we have two reversible Carnot-machines A and B, between a T_1 and a T_2 heat reservoirs. The parameters are as follows:

A is using ideal gas
A's efficiency $\eta_A = (T_1 - T_2) / T_1$.
A extracts heat Q_1 from the T_1 reservoir,
performs $L_1 = \eta_A Q_1$ work and loses $(1 - \eta_A) Q_1$ heat to the T_2 reservoir.

B is using a real gas.
B's efficiency is $\eta_B < (T_1 - T_2) / T_1$
B is operating in a reverse mode, in such a way that:
B is given $L_2 = \eta_B Q_1$ work and extracts $(1 - \eta_B) Q_1$ heat from the T_2 reservoir and
B loses $(1 - (1 - \eta_B)) Q_1 = Q_1$ heat to the T_1 reservoir

The total sum of the two engines is the following:

The T_2 reservoir loses heat
$[(1 - \eta_A) Q_1 - (1 - \eta_B) Q_1] = (\eta_B - \eta_A) Q_1 < 0$ due to $\eta_B < \eta_A$

and we have a total net gain of work $\eta_A Q_1 - \eta_B Q_1 = (\eta_A - \eta_B) Q_1 > 0$ (due to $\eta_B < \eta_A$

However, according to the second law of thermodynamics, it is impossible, therefore, out original assumption of $\eta_B < \eta_A$ is impossible as well. Q.e.d.

<u>To show that the second consequence is true,</u> let's assume that there is friction in the AB and CD segment of the Carnot cycle. In this case, the engine loses heat due to friction in both the AB and CD segment of the Carnot cycle. Therefore, the actual Q_1^* heat extracted from the T_1 reservoir: $Q_1^* < Q_1$ and the actual Q_2^* heat lost to the T_2 reservoir: $Q_2^* > Q_2$.

Consequently, $Q_2^* / Q_1^* > Q_2 / Q_1$, therefore the efficiency of the irreversible engine:

$$\eta^* = (Q_1^* - Q_2^*) / Q_1^* = 1 - Q_2^* / Q_1^*$$

is less then the efficiency of the reversible engine:

$$\eta = (Q_1 - Q_2) / Q_1 = 1 - Q_2 / Q_1$$

that is:

$$\eta^* < \eta$$

Due to these two consequences, we can state that for any gas used, reversible or irreversible processes:

$$Q_1/T_1 + Q_2/T_2 <= 0$$

Finally, it can be shown that any closed cycle can be approximated by a number of very small reversible and irreversible steps, and summing up (integrating) these steps gives us the following:

$$\oint dQ/T <= 0$$

where *the equal sign is valid only for totally reversible processes*.

One more fact needs to be shown before we can state Clausius's Entropy Law: integrating the dQ/T function along any path taken in the cycle from status A (p_A, V_A, T_A) to status B (p_B, V_B, T_B) will give us the same value. <u>The value of the integral does not depend on the actual path taken from A to B</u>.

It is obvious for reversible cycles because picking any two arbitrary points A and B on the whole cycle:

$$0 = (\text{rev}) \oint dQ/T = (\text{rev}_1) \int_A^B dQ/T + (\text{rev}_2) \int_B^A dQ/T =$$

$$= (\text{rev}_1) \int_A^B dQ/T - (\text{rev}_2) \int_A^B dQ/T$$

therefore

$$(\text{rev}_1) \int_A^B dQ/T = (\text{rev}_2) \int_A^B dQ/T$$

where rev_1 refers to one path leading from A to B and rev_2 refers to the other path leading from A to B.

Due to the above, we can define the **entropy function S(P)** for every point in the (V,p) continuum of the studied process, such that

$$S(P) = (\text{rev}) \int_{P_0}^{P} dQ/T$$

where P_0 is the status at absolute zero temperature where we assume that

$$S(P_0) = 0.$$

The Entropy function defined this way always belongs to one particular system (like a heat engine) that can move along the (V,p) continuum as it changes its status from one point (V_1, p_1) to another (V_2, p_2).

For irreversible processes we know from (9) that

$$(\text{irrev}) \int_A^B dQ/T + (\text{rev}) \int_B^A dQ/T < 0$$

therefore

$$(\text{irrev}) \int_A^B dQ/T < S(B) - S(A)$$

that is, for irreversible processes, the $(\text{irrev}) \int_A^B dQ/T$ integral is always less than the entropy difference between the two points A and B (which is calculated along any reversible path from A to B).

If we apply this result to a closed system that can not have any heat exchange with its environment, then dQ will be zero along the entire irreversible path from A to B, therefore we can see that for closed systems:

$$S(B) - S(A) > 0$$

that is: **for all irreversible processes of a closed system, the system's entropy can only increase**.

If we apply this statement to the whole universe, as a closed system, then it is obvious that the entropy of the universe will keep increasing until all processes stop in a final state of equilibrium. This state has been dubbed *"the heat death of the universe"* by overdramatic physicists who like drama just like everyone else.

The meaning of Entropy

When discussing the Carnot Cycle, we have deduced (from the ideal gas law) that during an **isothermal expansion** (same temperature) of the gas in the cylinder (against outside pressure), at temperature T, when increasing its volume from V_A to V_B, while its pressure drops from p_A to p_B, the gas extracts ΔQ heat from the boiler and performs the same amount of work:

$$L = \Delta Q = - \int_{V_A}^{V_B} pdV$$

Since, according to the combined gas law: $pV = NkT$ then $p = NkT (1/V)$ so the integral (according to basic calculus)

$$\Delta Q = NkT \ \ln (V_B / V_A)$$

The change in entropy during this expansion:

$$\Delta S = \Delta Q/T = Nk \ \ln (V_B / V_A)$$

Let's look at some probabilities. The probability of all the N molecules being in the volume V_B after the expansion is obviously one: $w_B = 1$

Now, the probability of all the N molecules being in the V_A ($<V_B$) volume (which is the situation before the expansion, so we can denote it with w_A, according to basic probability theory is:

If $N = 1$ then $w_A = V_A / V_B$
If $N = 2$ then $w_A = (V_A / V_B)^2$

and, generally for N molecules: $w_A = (V_A / V_B)^N$

(For example, throwing all 2-s with three dice has the probability of $(1 / 6)^3$)

So, we have:
$$w_B / w_A = (V_B / V_A)^N$$

and thus
$$\ln(w_B / w_A) = N \ln(V_B / V_A)$$

Looking at our equation of:

$$\Delta S = \Delta Q/T = Nk \ln (V_B / V_A)$$

finally we get:
$$\Delta S = k \ln(w_B / w_A)$$
that is:
$$S_B - S_A = k \ln w_B - k \ln w_A$$

which implies that

$$S = k \ln w$$

and this equation, which is called **Boltzmann's Equation**, means that the entropy of a system (S) is proportional to the natural logarithm of the probability (w) of the disorder in the energy distribution of the gas molecules (k is a proportionality factor).

Brownian Motion

Einstein and Marian von Smoluchowski (Polish physicist) tackled the problem (independently from each other, in 1905) by considering the diffusion of suspended particles from the point of view of the kinetic-theory of heat (at the time an unproven hypothesis).

Einstein's arguments are based on the conditions of dynamic equilibrium reached by a system of a large number of suspended molecules in a solution. In Einstein's words:

"We can look upon the dynamic equilibrium condition considered here as a superposition of two processes proceeding in opposite directions, namely: -

1. A movement of the suspended substance under the influence of the force K acting on each single suspended particle

2. A process of diffusion, which is to be looked upon as a result of the irregular movement of the particles produced by the thermal molecular movement"

In §3 of his paper, Einstein calculates the diffusion coefficient from this condition of dynamic equilibrium as:

$$D = (RT/N) \, 1/6\pi kP$$

where 'T' is the absolute temperature, 'N' the number of molecules in a mole of the suspended substance (Avogadro's number), 'k' is the coefficient of viscosity of the liquid and 'P' is the radius of the spherical molecule ('R' is the well-known coefficient from the ideal gas law).

In §4 of his paper, Einstein calculates the average of the square of one displacement component for the suspended molecules in time 't' seconds as:

$$\lambda_z{}^2 = <x^2> = 2Dt$$

From these two results, Einstein has the formula for the mean square displacement in time 't' as:

$$\lambda_z{}^2 = <x^2> = (RT/N)\,(1/3\pi kP)\ t$$

This displacement, as the formula shows, depends only on the temperature (T) and viscosity (k) of the suspension and the size (P) of the suspended particles, as well as the elapsed time (t).

Richard Feynman deduces the same formula (in a different form) from the basic Newtonian equation for motion under an external force (gravity) in a viscous liquid and the kinetic-theory definition of temperature ("$mv^2/2$ has a mean value of $\frac{1}{2}$ kT") (see "Lectures on Physics" II 41-8: "The Random walk").

Using different notations, Feynman came up with the formula of:

$$<R^2> = 6kT\ t\,/\,\mu$$

where $<R^2>$ is the mean displacement square, 'k' is the Boltzman constant, 'T' is the absolute temperature and 'μ' is the viscosity coefficient. His comments on the formula were:

"So one of the earliest determinations of the number of atoms [in a mole] was by the determination of how far a dirty little particle would move if we watched it patiently under a microscope for a certain length of time. And thus Boltzmann's constant k and the Avogadro number N_0 were determined…"

As I said before, within a few years Einstein's predictions were experimentally verified and the atomic theory was finally accepted by the scientific community.

Dreamers

Faraday was the first
who saw the invisible lines:
a force field permeating all of magnetic space…
…he didn't know he was proven right
by the Aurora Borealis;
he only knew that iron filings
sprinkled on a sheet of paper,
made, for him, the lines visible.
Many thought that it was risible
to imagine such things
if you couldn't see them, couldn't touch them,
how could they exist?

Einstein imagined space
filled with clocks ticking away,
at their own speed, in their own way.
Time flows at a different rate
for you and me if we don't wait
for each other but speed away
in opposite direction…
…no one could understand
his wild contention.

Sagan imagined the stars
we might visit someday
from our "pale blue dot"
and find new planets orbiting
in elliptical path,
some teeming with life, intelligence…
…then we'd know we are not alone,
that the whole universe
is our home.

All these dreamers shared this:
they were not afraid to dream,
to look beyond what may seem
rock-solid reality.
They knew there had to be
glorious variety in the cosmos:
in boundless infinity.

They are all dead now
but I still talk to them,
ask if I could share their vision…
…and they say that dreams are still free
and that's the way
it should, forever, be.

INDEX

Bibliography

Simon Singh: "Big Bang"
Carl Sagan: "Cosmos"
Arthur Koestler: "The Sleepwalkers"
Isaac Asimov: "Biographic Encyclopedia of Science and Technology"
Isaac Asimov's "Chronology of Science and Discovery"
John Ralston Saul: "The Doubter's Companion"
John Ralston Saul: "On Equilibrium"
John Horgan: "The End of Science"
Richard Feynman: "The Meaning of It All"
Richard Feynman: "Lectures on Physics"
Will Durant: "The life of Greece"
Will Durant: "The Story of Philosophy"
Morris Kline: "Mathematics and The Search For Knowledge"
James Gleick: "Isaac Newton"
Isaac Newton: "Principia"
Rene Taton: "History of Science"
J.L. Heilbron: "Electricity in the 17th and 18th century"
Giuliano Pancaldi: "Volta–Science and Culture in the Age of Enlightenment"
Immanuel Kant: "Critique of Pure Reason"
Bern Dibner: "Oersted"
James Hofmann: "André-Marie Ampère"
Oliver Darrigol: "Electrodynamics from Ampère to Einstein"
Dr Isaac Watts: "The Improvement of the Mind"
Pearce Williams: "Michael Faraday"
Basil Mahon: "The man who changed everything"
David Griffiths: "Introduction to Electrodynamics"
Herbert W. Meyer "History of Electricity and Magnetism"
Christian von Baeyer: "Maxwell's Demon"
Sheldon Glashow: "From Alchemy to Quarks"
G.I. Brown: "Scientist, Soldier, Statesman, Spy"
Osborne Reynolds: "Memoir of James Prescott Joule"
William Cropper: "Great Physicists"
S.P Thomson :"The Life of William Thomson, Baron Kelvin of Largs"
Roger Highfield and Paul Carter: "The Private Lives of Albert Einstein"
Albert Einstein: Annalen der Physik
Albert Einstein : "On the Electrodynamics of Moving Bodies"
Albert Einstein with Leopold Infeld: "The Evolution of Physics"

Richard Wolfson: "Simply Einstein"
A.P.French "Special Relativity"
Brian Greene: "The Fabric of the Cosmos"
Carl Sagan: "Pale Blue Dot"
Lawrence M. Krauss: "The Physics of Star Trek"
Isaac Asimov: "Extraterrestrial Civilizations"
Richard Rodes: "The Making of the Atomic Bomb"
Andrei Sakharov: "Memoirs"
Leonard Susskind and George Hrabovsky: "The Theoretical Minimum"
Nick Herbert: "Quantum Reality" (1985)
Richard Panek "The 4% Universe"
Lisa Randall "Knocking on Heaven's Door";
Lisa Randall "Warped Passages"
George Gamow "A Star Called the Sun"
Amir D. Aczel "Entanglement"
Hugh Aldersey-Williams "Periodic Tales"
Richard P. Feynman "QED - The Strange Theory of Light and Matter"
Robin McKown "Mendeleyev and his Periodic Table"
Ruth Moore "Niels Bohr"
David Wilson "Rutherford – Simple Genius"
Isaac Asimov "The Search for the Elements"
Michael White and John Gribbin "Stephen Hawking – A Life in Science"
Richard Feynman "Surely You're Joking, Mr. Feynman!"
Richard Feynman "What Do You Care What Other People Think?
Eve Curie "Madame Curie"
John Gribbin "Richard Feynman"
Suzan Zannos "Dimitri Mendeleev"
Steven Weinberg "Dreams of a Final Theory"
Sam Kean "The Disappearing Spoon"
P.W.Atkins "The Periodic Kingdom"
J.L.Heilbron "The Dilemmas of an Upright Man"
Theodore Gray "Elements"
Zeilik, Gregory, Smith "Introductory Astronomy & Astrophysics"
David Lindley: The End of Physics (1993)
Helge Kragh: "The Quantum Generations". (1999)
Lee Smolin: "Trouble with Physics" (2006)
Sheilla Jones: "The Quantum Ten" (2008)
Alexander Unzicker and Sheila Jones: " Bankrupting Physics" (2013)

About the author

Francis Mont has been living in Canada for the past 50 years after he emigrated from his native Hungary where he studied science and received a degree in Theoretical Physics. Over the years he did research, application, and teaching in Mathematics, Physics, and Computer Science. He is interested in profound questions, both in science and in social philosophy. He is a 'big picture' person, focusing on fundamental principles and the defining essence of the topic at hand. He also pursues independence and self-reliance to the best of his abilities, as his solar power system and year-round greenhouse demonstrate. He writes poetry, plays classical violin, dabbles at wood carving, and has not yet stopped building the house he and his wife and (currently) five cats live in.

Ordering Information

You can order a copy of this book at the following venues:

- www.amazon.ca
- www.amazon.com
- www.alibris.com
- www.abe.com
- www.montland.ca

or by sending email to the author to the following address:

books@montland.ca

I will respond to queries within 24 hours.

www.ingramcontent.com/pod-product-compliance
Lightning Source LLC
Chambersburg PA
CBHW081457200326
41518CB00015B/2285